FRONTIERS OF APPLIED MATHEMATICS

FRONTIERS OF APPLIED MATHEMATICS

Proceedings of the 2nd International Symposium

Beijing, China 8 – 9 June 2006

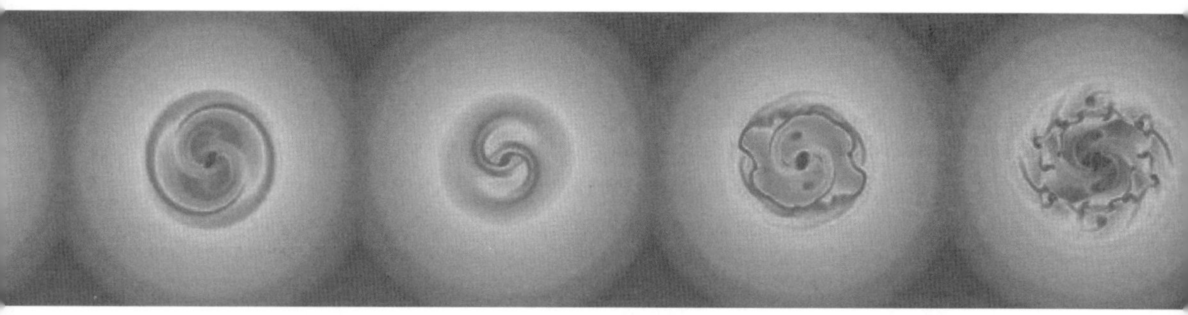

editors

Din-Yu Hsieh • Meirong Zhang • Weitao Sun

Tsinghua University, China

World Scientific

NEW JERSEY • LONDON • SINGAPORE • BEIJING • SHANGHAI • HONG KONG • TAIPEI • CHENNAI

Published by

World Scientific Publishing Co. Pte. Ltd.
5 Toh Tuck Link, Singapore 596224
USA office: 27 Warren Street, Suite 401-402, Hackensack, NJ 07601
UK office: 57 Shelton Street, Covent Garden, London WC2H 9HE

British Library Cataloguing-in-Publication Data
A catalogue record for this book is available from the British Library.

FRONTIERS OF APPLIED MATHEMATICS
Proceedings of the 2nd International Symposium

ISBN 13 978-981-270-456-6
ISBN 10 981-270-456-6

Printed in Singapore by World Scientific Printers (S) Pte Ltd

PREFACE

The Second International Symposium on the Frontier of Applied Mathematics was held at Tsinghua University on 8-9 June 2006. It was also an occasion to celebrate the 90th Birthday of Prof. Chia-Chiao Lin (C.C. Lin) and to honor his contributions to the advancement of applied mathematics. This volume is a collection of 14 original articles by world famous scientists from various parts of the world, including China, United State of America, Japan and Italy. The symposium presented a special opportunity to discuss state-of-the-art researches in applied mathematics.

Applied mathematics play an important role in all fundamental sciences and engineering applications. The scope of applied mathematics is very broad, and can be best described by the following words of C.C. Lin:

"The principal theme is the interdependence of mathematics and the sciences. In common with the pure mathematician, the applied mathematician is interested in the stimulation of the development of new mathematics, - but with primary emphasis on those aspects directly or at least very strongly motivated by scientific problems. In common with theoretical scientists, the applied mathematician seeks knowledge and understanding of scientific facts and real world phenomena through the use of mathematical methods."

This volume provides an introduction to frontier research of certain areas of applied mathematics. The scope of this proceeding also coincides with the past and present research interests of Professor Lin. The 14 articles in this proceeding can be essentially divided into four parts: **neural science, protein structure, astrophysics, and nonlinear waves.**

In closing, we would like to thank all authors for submitting their works. They are Nancy Kopell, Din-Yu Hsieh, Frederic Y.M. Wan, Kerson Huang, S. Takahashi, Haijun Zhou, Chi Yuan, Giuseppe Bertin, Chung-Pei Ma, Frank H. Shu, Mark J. Ablowitz, Michael I. Weinstein, David J. Benney, Jianke Yang. We also thank Douglas N. C. Lin for his interesting lecture in this symposium. The preparation and organization of the symposium, as well the preparation of this volume, are supported by the staffs of the Zhou Pei-Yuan Center for Applied Mathematics at Tsinghua University (ZCAM).

CONTENTS

MULTIPLE RHYTHMS AND SWITCHES IN THE NERVOUS SYSTEM

N. KOPELL

Department of Mathematics and Statistics, Boston University, Boston MA 02215

D. PERVOUCHINE

Center for BioDynamics, Boston University, Boston MA 02215

H. G. ROTSTEIN

Department of Mathematics and Statistics, Boston University, Boston MA 02215

T. NETOFF

Department of Biomedical Engineering, Boston University, Boston MA 02215

M. WHITTINGTON

School of neurology, Neurobiology and psychiatry, University of Newcastle, Newcastle upon Tyne NE2 4HH, UK

T. GLOVELI

Institute of Neurophysiology, Charite-University Medicine, Berlin 10117, Germany

Networks of neurons in the nervous system can produce a variety of temporal patterns of different frequency; the same network can produce different rhythms at different times, or several rhythms at one time. We focus here on the gamma (30-90 Hz) and theta (4-12 Hz) rhythms produced in the hippocampus, a part of the nervous system critical for learning and recall. We discuss experiments and models that suggest that there are separate subnetworks that produce the different rhythms; the sharing of components of the networks induces competition between the rhythms, which can lead to suppression of one of the rhythms, or nesting of the rhythms. We show how low-dimensional maps can help to understand the properties of the cells and networks to allow this to happen.

Keywords: neural dynamics, gamma rhythm, theta rhythm, hippocampus, low-dimensional maps

1. Rhythmic dynamics in the nervous system

The nervous system produces dynamics at all states of wake and sleep. These can be detected via EEG and MEG measurements in non-invasive ways, and through invasive electrophysiology techniques. All these techniques indirectly record the activity of electrical currents that are created by neurons in the brain, signaling to other neurons. These electrical and magnetic signals can be analyzed for their spectral content, and it has been found that the spectral content varies with the cognitive state of the subject: certain frequency ranges, such as the gamma frequency (roughly 30-90 Hz) are associated with attention, active processing of early sensory signals, short term memory and other behavioral situations. The theta rhythm (roughly 4-12 Hz in rats and 4-8 Hz in humans) is seen in active exploration in rats, and is believed by many to be important for the recording and feedback of memory, especially memory of sequences of places or events.

The mathematical questions associated with these rhythms concern their biophysical origins, and how the same bits of tissue can, in different behavioral situations, produce different rhythms or combination of rhythms. By piecing together the biophysical substrate of the rhythms, one is then in a position to start investigating how the networks can process different kinds of spatially and temporally patterned signals; such information is critical to understanding how the brain makes use of these dynamics for cognitive function.

Many of the rhythms that are seen in behaving animals have been replicated *in vitro* (in slices of brain tissue). This allows the use of much more powerful measures to discover the biophysical nature of the different rhythms. Much of the work that has been done deals with single spectral bands. However, recently, some labs have been able to reproduce a combination of rhythms that appear very much as they do *in vivo*. This gives us a window into how the nervous system may be using interactions of rhythms to process information. One of the most widespread pair of interacting rhythms is the gamma/theta pair (Chrobak *et al.* , 2000), and that is the focus of this talk.

2. Gamma, Theta and in vitro preparations

In vitro preparations are different from one part of the nervous system to another, and the anatomical focus of this talk is on a part of the hippocampus known as CA3. This is a section of the so-called “hippocampal loop”, in which signals come from the neocortex, go around this loop (with some further possible inputs and outputs), and then go back to the neocortex, presumably in a different form. This loop is believed to be crucial in the encoding of new memories. The CA3 region is known *in vivo* to produce both theta and gamma in a nested manner. In order to address what the system might be doing with those rhythms, it is helpful to understand what creates them, and how they related to one another. A recent breakthrough in this came from the work of Gloveli *et al.* (2005) who showed that a network in a slice that produces both can in some sense be parsed into interacting and overlapping gamma and theta networks. The idea, though not the work, was simple: By slicing in the standard transverse direction, one gets a gamma rhythm, in the longitudinal direction a theta rhythm, and in a direction in between, a nested rhythm. The modeling and mathematical issues addressed in this paper concern how this comes about. The essential idea is that the full network is composed of different kinds of cells, and different subnetworks of those cells are responsible for different rhythms. Under different conditions — either the angle of the slice or modulatory changes in the chemical soup — different subnetworks can take over and produce different rhythms. We believe this is a general principle in the nervous system. We first describe mechanisms separately for the gamma and theta rhythm, then return to this example to talk about how they interact in a larger network.

3. The Mathware: Voltage-gated conductance equations

The simulations we discuss come mainly from the so-called Hodgkin-Huxley equations or, more generally, "voltage-gated conductance equations" (Dayan & Abbott, 2005). These are equations for a single neuron, or a network of neurons. For each cell, the main equation is for the voltage difference across the cell membrane, and the equation represents conservation of currents passing through the membrane, balanced by a capacitance current. Each of the currents is generated by single kind of ion (or combination of ions) that pass through molecular "channels" that open and close with voltage-dependent kinetics. For each cell, the current conservation equations are supplemented by other equations that describe the opening and closing of channels as the voltage changes. The full equations are PDEs, describing the voltage across the spatially extended cell. However, in simplified models, the spatial extension is often ignored, giving rise to a set of equations known as a "point" neuron. In this paper, most simulations use point neurons. The main equation is:

$$C\frac{\partial v}{\partial t} = -I_{ion} + \nabla^2 v - \sum I_{synapse}$$

where

$$I_{ion} = gm^j h \cdot (v - V_R).$$

Here each ionic current is given by Ohm's law: it is a product of the conductance $gm^j h$, which is the inverse of the resistance, and a driving force $v - V_R$. The latter acts like a battery with a "reversal potential" that depends on the kind of ion; g is a constant, and m and h are gating variables, describing channels in the membrane that open an close in a voltage dependent manner, according to differential equations of a form

$$\frac{dx}{dt} = (x_\infty(v) - x)/\tau(v)$$

The functions $x_\infty(v)$ are sigmoid as in figure 1.

Each neuron may have a different set of conductances and different types of neurons generally do have such differences. Equations for a network of neurons consists of separate sets of equations for each cell, coupled via the voltage equation with other currents labeled $I_{synapse}$. The coupling currents can be excitatory or inhibitory; roughly speaking, excitatory currents increase the voltage, making it easier for the receiving (post-synaptic) cell to fire, while inhibitory currents make it harder for the post-synaptic cell to fire.

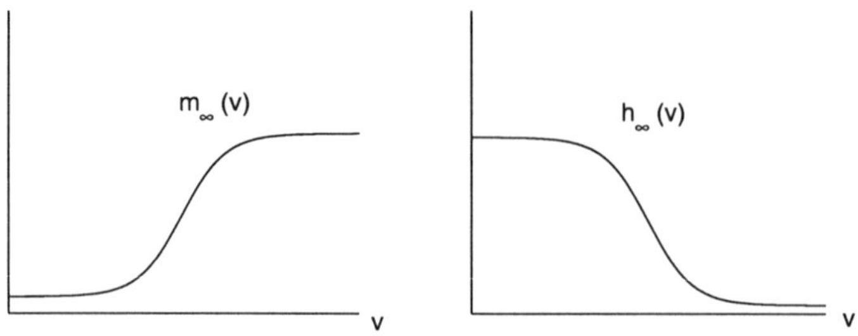

Fig. 1. Activation curves for gating variables. The vertical axis is the fraction of channels open when the voltage is held at a given value. Such a curve is usually monotonic, and can either increase or decrease with voltage.

4. Some mechanisms for the gamma rhythm

There are (at least) several different biophysical mechanisms that produce a gamma rhythm. Two are discussed here. Only one is directly important for the CA3 story, but the other provides a contrast to another phenomenon that is central to the story.

Unlike some other rhythms, gamma is very much an inhibition- based rhythm (Whittington *et al.* , 2000). The reason comes basically from how common inhibition acts to synchronize target cells. It is illustrated by the simplest form of neural equations, the "integrate-and-fire" neuron, with an additional term for inhibition.

$$\frac{\partial v_1}{\partial t} = I - v_1 - g_{syn} e^{-t/\tau}$$

$$\frac{\partial v_2}{\partial t} = I - v_2 - g_{syn} e^{-t/\tau}$$

The inhibition is modeled here by a synapse that turns on instantaneously, and then decays exponentially with time constant τ. The essential point is that the time it takes inhibition to wear off is the longest time constant. The inhibition provides a kind of quasi-steady state that is tracked by the voltage of each of the target cells. By the time the inhibition wears off, the cells are essentially at the same voltage, and will fire synchronously, provided the target cells are identical. Thus initial conditions are wiped out by the inhibition.

If the cells are not identical, this mechanism gives phase differences for the firing times of the target cells. If there are other currents involved, as in other rhythms, the effect of the synapses does not necessarily lead to synchronization. A similar mechanism, though harder to understand mathematically, produces a gamma rhythm for a coupled system of equations: the common inhibition here comes from the population (White *et al.* , 1998; Chow *et al.* , 1998; Wang & Buzsaki., 1996). This is called ING or inhibitory network gamma. We will contrast it later with another network in which the connections are inhibitory but the cells don't synchronize.

The mechanism for gamma is actually simpler - in principle - when there are excitatory cells involved. Pyramidal Interneuron Network Gamma (PING) is induced *in vitro* by tetanic stimulation of tissue: the stimulation is higher frequency than gamma; when it is over, the network keeps firing at gamma for a short time, with the excitatory pyramidal cells (E-cells) firing on each cycle, as do the inhibitory fast-spiking interneurons (Whittington *et al.* , 2000). The PING rhythm is easiest to understand when there is one excitatory and one inhibitory cell (I-cell). The excitation from the pyramid (excitatory cell) causes the I-cells to spike, which inhibits both cells, and the cycle begins again when the inhibition wears off. We think of this as "ping-pong", because the action bounces between the two kinds of cells. The only important currents other than synaptic currents are the standard spiking currents.

Things get more subtle when there is a larger population. One of the subtle aspects is that the synchronization can happen even in extremely sparse and heterogeneous networks, as shown and explained mathematically by Börgers and Kopell (Börgers & Kopell, 2003; Börgers & Kopell., 2005). Here it is the I-cells that synchronize their target population, the E-cells, as described above. The E-cells synchronize the I-cells more crudely, but enough to add to the process.

This mechanism also depends on the time scale of decay of inhibition, but not as much (Whittington *et al.* , 2000). If there are both I-I and E-I connection in the network, changes of parameters can take the network between the ING and PING regimes; in the PING regime, the I-I connections (or even E-E connections) are essentially irrelevant. This has implications for responses to heterogeneity and noise.

5. A theta-rhythmic cell

Gamma is the simplest rhythm because it requires no intrinsic currents other than those that produce the spike currents; it is essentially a network rhythm, in which inhibition is critical. By contrast, the theta rhythm is produced by many different kinds of cells in the nervous system. One of these in the area we are focusing on is called the O-LM cell, short for oriens-lacunosum moleculare, for the latin names of the layer of the cell body (oriens) and where it projects to (figure 2). We will refer to it as an O-cell. This single-cell rhythm depends on other intrinsic currents. One of these, which plays a large role in the network behavior, is called the h-current, for heart, where it was found (it is also called "sag current", "weird current", "anomalous rectifier").

The unexpected properties of this current come from its interaction with excitation and inhibition. The interactions have to do with how intrinsic conductances depend on the voltage of the cell. For most conductances, increasing the voltage increases the conductance, i.e., opens the gate more. However, there are some con-

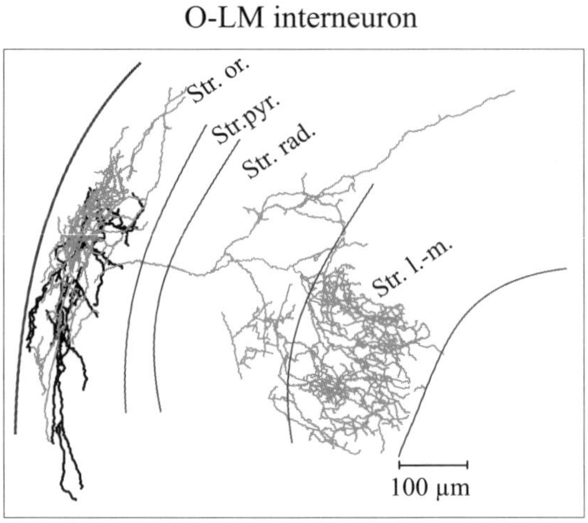

Fig. 2. Reconstruction of an O-LM cell. The dendrites are in str. oriens, and the axons branch into str. lacunosum moleculare, as well as str. oriens.

ductances that act in an anti-intuitive manner: the higher the voltage, the smaller the conductance. This is true of a class of currents known as "hyperpolarization activated currents", which includes the h-current. These currents turn up in many cells in the nervous system, including the O-cell. Indeed, the h-current is the main current that determines the voltage between spikes for those cells (Saraga *et al.* , 2003; Rotstein *et al.* , 2005).

When inhibition or excitation is added to a cell that has nonlinear conductances between spikes, it doesn't just change the voltage by adding a new current; it changes the other currents that are sensitive to the voltage. Adding inhibition to the O-cell can initially lower the voltage, but then the h-current turns on and makes the voltage go back up. So inhibition in such a cell can actually make the next spike come faster!

To get a rhythm, cells must fire in a coherent way; one of the mathematical themes of this paper concerns synchronization properties of cells coupled by synapses. As we discuss more below, the h-current has massive effects on network behavior: This current totally reverses the synchronizing properties of excitation and inhibition (Crook *et al.* , 1998; Ermentrout *et al.* , 2001; Netoff *et al.* , 2005; Acker *et al.*, 2003). Biophysical O-cells are inhibitory, and model O cells, when connected with inhibition, do not synchronize for most initial conditions (unlike the more standard inhibitory cells described above); this can be traced to the effects of inhibition on this current. We discuss below how the properties of this current can get built into low-dimensional maps that capture the behavior.

6. Structure of Hodgkin-Huxley equations and low-dimensional maps

The Hodgkin-Huxley equations have structure that sometimes enables one to use low-dimensional approximations, at least near some set of relevant trajectories (Clewley *et al.* , 2005; Pervouchine *et al.* , 2006; Rotstein *et al.* , 2006). For that, we have to recall that the main equation is a conservation of charge, with a sum of different currents. In different voltage ranges, some of these currents are not active, and then the kinetic equations that go with their gating variables are not relevant. Within the active set of currents, some of the gating variables may have time scales that are very long, and hence the relevant variable is essentially locally constant, or very short and hence the variable is slaved to other slower variables. Some kinetics are themselves voltage dependent; gates can be slow to open, but very fast to reset to closed once the voltage gets sufficiently large. That means they reset after a cell spikes, again lowering the number of really independent variables. We use all of these in constructing low-dimensional maps of interacting cells.

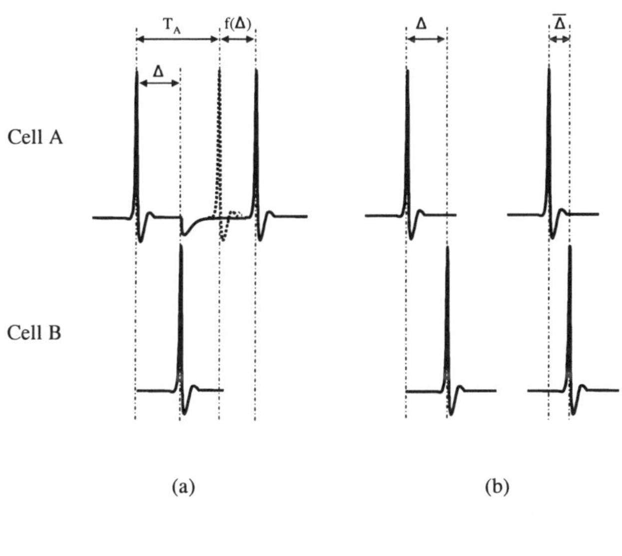

Fig. 3. Spike time response method. (a) Construction of the spike-time response curve. (b) Construction of the spike-time difference map.

Figure 3 gives the basics of the idea of the lowest possible dimensional maps in question. The first panel shows the idea of the Spike-Time Response Function, which is essentially the same as a phase response curve, only explicitly in time rather than phase. Cell B sends a pulse to cell A at some time Δ after a spike of A, and the STRC measures the change in the time A next spikes compared to when it would have spiked. In the figure, T_A is the time it would have spiked, and the difference is called $f(\Delta)$. For this to be at all useful, it has to be checked that no other variable besides Δ - e.g., the status of any of the gating variables - makes a difference to the time of the next spike. Though we will not discuss that, where we use such formalism, such assumptions generally follow from the reduction ideas sketched above.

Now we consider what happens when two cells interact, and ask if the two cells lock at some phase. We construct a map that takes the time difference between the spikes of the cells on one cycle to the time difference after they have both spiked. There is an extra hypothesis that goes into this: we have to know the order of the spikes to do this construction. In the figure given in panel b, we are assuming that the spikes do not change order. One gets the map by following the effects of each spike on the other cell, starting from a fixed time difference Δ.

The significance of the STRCs and STDMs are that they are the bridges between the biophysics of the cells and their synapses and the behavior of the network. Changing anything biophysical leads to changes in these functions and maps, and allows one to understand how altered biophysics can change the network behavior.

7. Where does the theta come from?

We said above that a pair of O-cells, coupled by its own inhibition, does not synchronize. We can understand this from constructing STRCs and STDMs corresponding to the interacting O-cells. The first panel in figure 4 gives an example of STRCs for an O-cell getting input from another O-cell. Both the receiving cell and the input is different from that of a simple I-cell (which is like an integrate-and-fire cell) getting I-input. The receiving cell is different because it has the special currents mentioned (especially the h-current). The input is different because the current from an O-cell lasts several times longer than that from an I-cell. Both of these differences contribute to differences in the STRC compared with that of I cells input to I-cell.

The main thing to notice about the STRC is that an input that arrives soon enough after the receiving cell spikes actually speeds up the next spike. This is the effect of the h-current mentioned before (Dickson *et al.* , 2000). Later in the cycle, the inhibitory input slows down the next spike. This is shown for different values of the maximal h-current. Note that the larger the h-current, the more of an initial effect there is. One can reason from the kinetics and voltage dependences of the h-current why the STRC looks the way it does, but we do not discuss that here. The second panel (figure 4) is a measurement of STRC using what is known as a "dynamic clamp": A computer acts as one cell, and feeds in the synaptic current at wanted times, while controlling the frequency of the receiving cell (Netoff *et al.* ,

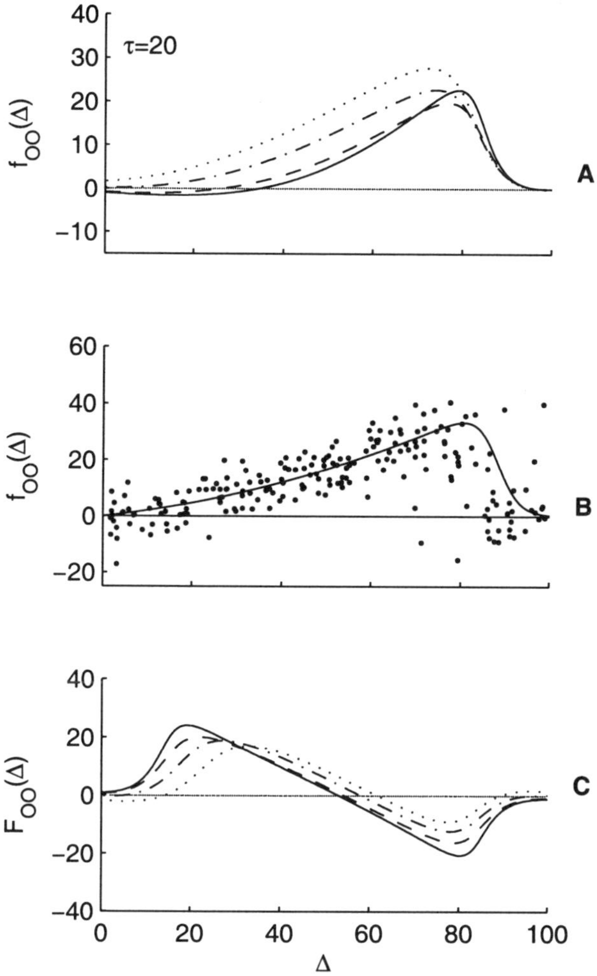

Fig. 4. STRC and STDM for O-O network. (A) STRCs for different levels of the h-current. The solid line corresponds to the highest level of h-current. (B) An experimentally determined STRC. (C) The STDMs constructed from the STRCs in panel A.

2005). Note that it has the same shape as the model STRC, while being considerably noisier.

The next panel (figure 4) gives the associated spike time difference map, also for several values of the h-current. What is shown is not the map itself, but the difference F between the map and the identity. Thus, a fixed point for the map is a zero for the function F. Recall that a fixed point for a map is stable if the derivative is between -1 and + 1; this implies that the derivative of F must be between -2 and 0. The figure shows that there is such stability for a point somewhere in the middle. One cannot tell from this figure, but from symmetry one can reason that this point corresponds to antiphase; the period, which is not the same in the coupled network as in the individual cell, is twice the value of the time at the fixed point of the map.

The punch line of the above is that O-cells, when coupled by their natural coupling, do not synchronize. Hence, it is not possible to get a coherent theta rhythm from a set of O-cells even though each cell is capable of firing at theta; the population rhythm is faster. The same is true for a large number of O-cells. This is in direct contrast to a pair of I-cells, which does synchronize. So, we have a mystery: where does the population theta rhythm come from?

Earlier, we explained how common inhibition can synchronize cells. We also said then that this depends on the cells receiving the inhibition being simple enough. There are I-cells in the CA3 network that produces the theta and one might think that the common inhibition might synchronize the O-cells. But it doesn't necessarily do that. If the I-cell input has the same rough frequency as the O-cells, then it does synchronize them. However, if the input is much faster than the natural frequency of the O-cells, then something else happens, as shown above. Each time there is an inhibitory input, the h-current increases; this partially entrains the O-cells, but cannot make those cells fire much faster than they want to. Instead, the O-cells miss many I-cell cycles, but do fire at a specific phase when they fire. This doesn't synchronize the O-cells because they can skip different cycles (the R. H. panels in figure 5 show the behavior of the h-currents for the two O-cells). Thus, there is not synchronization at the theta frequency.

However, there IS a way to get theta in an O/I network that is not intuitively obvious. One needs both the common inhibition to the O-cells and feedback from the O-LM cells to the I-cell. This is un-intuitive, since I-cells can synchronize, but not at theta, and O-LM cells cannot synchronize at any frequency. Figure 6 shows the synchronization. It is critical that the decay time of the O-cells is much longer than that of the I-cells: The long decay time groups the spikes of the I-cell(s) into small bursts. So even when the I-cell is driven enough to fire at high frequencies, one gets the theta rhythm. If there are multiple I-cells so I-cells also get I-inhibition, can get theta and gamma here, nested.

How this works is not intuitive and working with low-dimensional maps helps to explain this. To understand this a little better, we start with just one I-cell and one O-cell. Unlike the O-O circuit described earlier, this is not symmetric. So we don't expect to get synchrony or antiphase, and indeed we don't. The first question is only about whether they lock at all, and at what relative phase.

For this, we can use the techniques described before, using a Spike-Time Response Curves. The first two panels in figure 7 below show those functions, which measure the effects of I on O and O on I. The effects of I on O depend on the h-

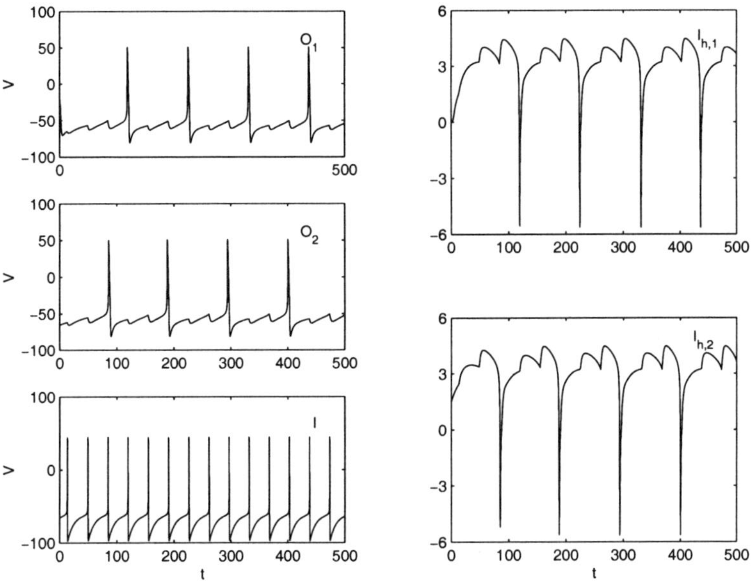

Fig. 5. O-cells need not synchronize with common inhibition. Left panel: The voltages of a pair of O-cells and an I-cell. Right panel: The conductances of the h-current for the two O-cells.

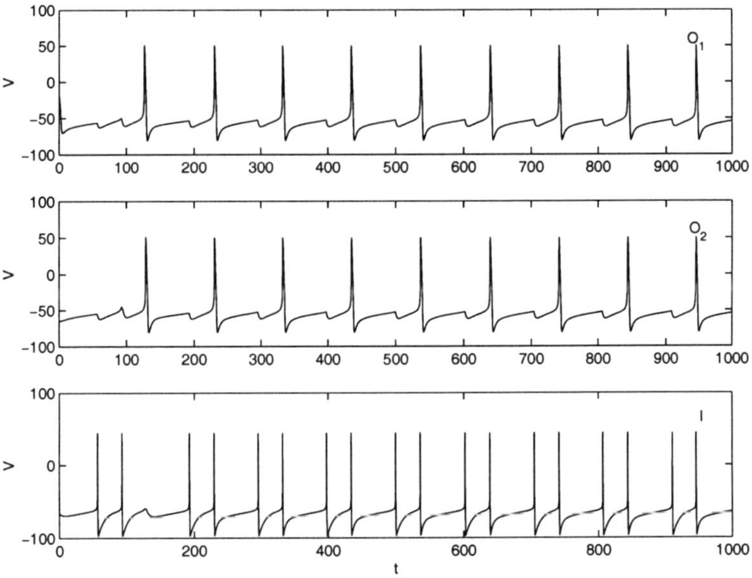

Fig. 6. Adding feedback to the I-cell achieves synchrony of the O-cells. Voltages for the two I-cells and the O-cell.

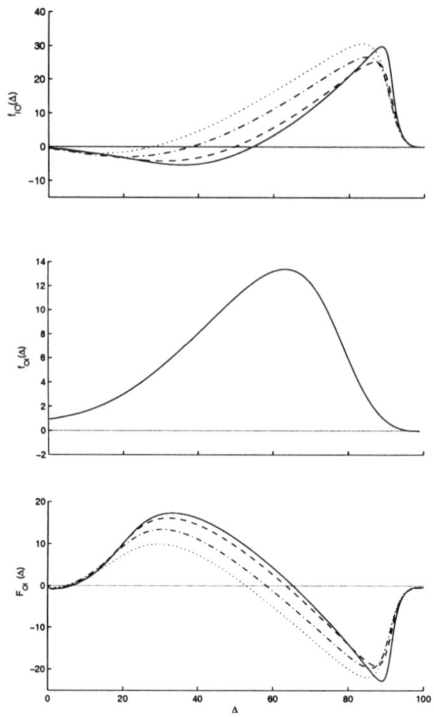

Fig. 7. STRCs and STDM for the O-I network. Top panel: Effect of the inhibition from the I-cell onto the O-cell, at different levels of h-current in the O-cell. Middle panel: Effect of inhibition from the O-cell onto the I-cell. Bottom: STDM giving the difference in time between the firing of the O-cell and the next firing of an I-cell as a function of that difference in the previous cycle.

current in O-cell, so there are several curves. The last panel (figure 7) is the STDM minus the identity, as before. The fixed point of the map, corresponding to the zero of this curve is about 50-70 ms into the cycle, and represents when the I-cell fires after the O-cell in steady state. From the biophysics, we can understand this: The I-cell fires, primes O-cell to fire shortly after the I-cell inhib wears off. The I-cell can fire only after the O-cell inhibition wears off. Both h-current and differences in the decay times lead the interval between I and O to be considerably shorter than the O to I interval. In spite of the fact that there are many variables in the relevant equations, a 1-D map is accurate because the variables are slaved to the spike times, even the h-current, which resets whenever the O-cell fires.

One might ask what kind of understanding one gets by thinking about maps, beyond just simulating and seeing what happens. The answer is that it helps us understand what matters to the synchronization of the O-cells, in particular that the long kinetics of the O-cell-derived inhibition is critical.

Once we have more than one O-cell, we cannot use 1-D maps - there is more than one real degree of freedom, not slaved to one spike time difference. With two O-cells, the system can be described by a 2-D map, with variables Δ, the difference between the O-cell spikes, and σ, the difference from the second O-cell spike to the I-cell spike. This can be analyzed as a 2-D map, but the essential insight comes from looking at slices of that map gotten by fixing σ. Figure 8 shows several such slices. Unlike our other graphs, where we are interested in only fixed points, we now draw this as a map, i.e., without taking away the identity, so we can use standard cobwebbing (a graphical method for understanding the dynamics of iterated maps) to see the behavior.

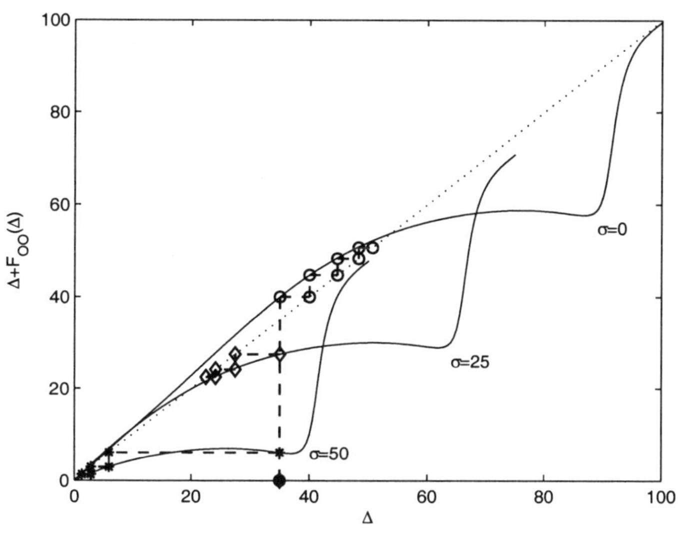

Fig. 8. Slices of 2-D map for fixed σ

The lowest graph is for $\sigma = 50$, and we see that the Δ goes to zero and stays there by the first cycle. By contrast, the ones with a much smaller σ do not go to zero: for $\sigma = 25$, the Δ stays near a much higher value for many cycles, and for $\sigma = 0$, the O-cells actually move further apart. Though the slices do not capture the full behavior of the 2-D map, they do accurately measure steady-state behavior, and this shows that, unless the σ is large enough, one does not get synchrony of the O-cells as a steady state. The value of σ is directly tied to the decay time of the O-cell inhibition, and decreases when the latter decreases. Thus, it is necessary to have the O-cell decay time long enough to make the whole network produce theta when there is more than one O-cell.

8. All together now: a larger network

We now return to the original question: why is it that when CA3 is cut into slices of different orientation, one sees different rhythms? The clue to this comes from the anatomical work that was done by Gloveli and collaborators. The key observation is the way that basket cells (I-cells) and O-cells are arranged. The basket cells tend to arborize more in the plane of the standard transverse slice, the one that gives the gamma rhythm; the O-cells, by contrast, tend to arborize in clumps along the longitudinal axis. In the transverse plane, there is only one clump, but in the longitudinal plane, there are 2-3 separate clumps.

Our question: could this be enough to account for the differences in rhythms? This is an obvious modeling question. We considered building a 3-D simulation to test this, but we believe that dynamical structures are most easily understood in "minimal" models. So what IS minimal here? Our guess, which our simulations support, is that what matters is the relative strengths of the O- and I-cell projections on the excitatory neurons; we could test this in a model that did not have much, if any, spatial structure. The model network is shown in figure 9. It has a pair of O-cells, a pair of I-cells and a single E-cell that has a soma and a dendrite. We felt that the 2-compartments were important, since the I-cells project to the soma and the O-cells to the ends of the dendrite. Within this model, it was possible to alter the relative strengths of the synapses from the O- and I-cells.

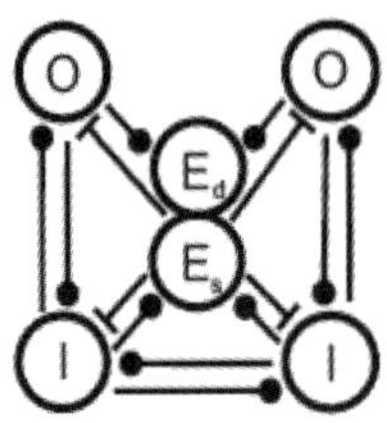

Fig. 9. Schematic of model for nesting of gamma and theta rhythms.

The simulations correspond to the transverse, longitudinal and coronal slices in terms of strengths of connections. The first panel (A) in figure 10 shows the E-cell soma producing a gamma rhythm interacting with the I-cells (which are synchronous). This is a standard PING. The O-cells, which are not synchronous, fire at a lower frequency, producing a small theta envelope to the amplitude of the synaptic currents, as seen in the experiment. To get this rhythm, the parameters are set so that the I to E connections are relatively strong, as are the I-O connections. The O-cells are firing at roughly theta frequency because the h-current builds up in the cells over several gamma cycles, due to the I-cell inhibition, as understood from the interaction between the I and O cells. As discussed before, the common inhibition need not synchronize the O-cells. This is also seen in the experimental data.

To get the theta oscillations shown next, we model the input from the extra clumps of O-cells as increasing the input from the O-cells to the I- and E-cells, and decreasing the input from the I-cells (Panel B). This increasing inhibition from the O-cells, with its much longer decay time, changes the firing of the E-cells to much slower, which then contributes to slowing down the I-cells. The critical switch in behavior is that the I-cells now fire at theta, not gamma. The interaction of the I- and O-cells produces the almost synchrony of the O-cells; if the O-cells had been identical, they would be synchronous, but to match the data, which is more ragged, the cells are not identical in their drives or inputs from the I-cell. The order of the firing - O-cells, I-cell and E-cells, match that in the data. Panel C is the simulation of the coronal slice, in which there is both gamma and theta in the excitatory cells, which are the output cells of the network. There is now a nesting of the gamma and theta, as seen in the experimental data, with the longer period that of theta, and the frequencies in each cluster of spikes gamma. That is achieved just by using intermediate parameters. Recall that one obtains multiple I-spikes for each O-spike provided that the I-cells are sufficiently excited. This is compatible with getting only one spike, as in panel B, when the O-I inhibition is strong, but when that is decreased, the nested rhythm can pop up.

9. Multiple rhythms, switches and bursting

The last set of simulations (panel C), with the nesting gamma and theta, is very reminiscent of bursting systems. In its simplest form, bursting comes from a single neuron in which there is a structure of a reduced H-H equation in 2 dimensions, plus a third variable which is much slower. As the third variable, thought of as a parameter, changes, the 2-D system can change between having a critical point and a limit cycle (Izhikevich, 2006).

The system we are considering has many more variables than 3, and minimally 3 different cells (one O, one I and one E). However, using reduction of dimension ideas, the effective dimension at any given time is actually much lower. Furthermore the behavior is very similar to classic bursting: the E/I network produces the gamma oscillation when the inhibition from the O-cell, which is slower to decay, is low enough to allow it. Thus, we hypothesize that the switching behavior associated with the three different kinds of behavior shown in figure 10 can be understood in a way similar to that of bursting, using reduction of dimension ideas. This is currently in progress.

10. Discussion

The gamma and theta rhythms are present together in various parts of the nervous system under various behavioral situations. We focus on a slice preparation that can produce either gamma or theta or a combination of them both. The same network

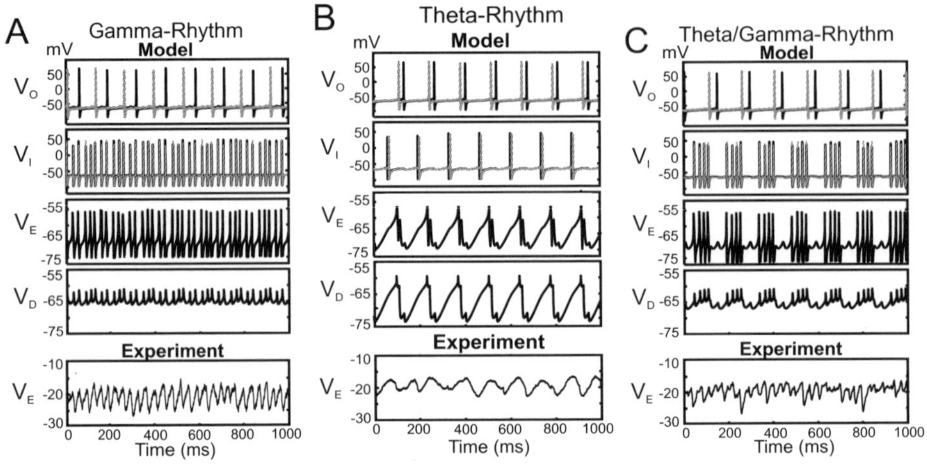

Fig. 10. Simulations of model. (A) the I- and E-cells produce a gamma rhythm. (B) All cells produce a theta rhythm. (C) Gamma rhythm nested inside a theta rhythm.

is involved in all three situations; changes of parameters change which subnetworks control the behavior of the full network. For gamma, the centrally important subnetwork is the E/I one; for theta it is the O/I interaction. By changing parameters one can get different subnetworks to dominate the rhythm. The interaction of the rhythms is reminiscent of bursting, in which the state of the fast variables depends on the state of some slower one.

The key functional issue is how the theta oscillation might be useful for coordinating temporal sequences of gamma-induced cell assemblies. It is known that gamma rhythms are local, while theta coherence is more global (Gloveli *et al.* , 2005). It is also known (Jalics, Kispersky and Kopell, unpublished observation) that theta interacts with gamma, and theta-frequency inputs can change the phase of gamma rhythms. The understanding of the biophysical bases of gamma and theta rhythms provides clues for how the networks react to input that is spatially as well as temporally structured.

References

1. Acker, C., Kopell, N., & White, J. 2003. *J. Comp.Neurosci*, **15**, 71.
2. Börgers, C, & Kopell, N. 2003. *neurocomp*, **15**(3), 509–538.
3. Börgers, C., & Kopell., N. 2005. *Neural Computation*, **3**, 557.
4. Chow, C., White, J., Ritt, J., , & Kopell, N. 1998. *J. Comput. Neurosci.*, **5**, 407.
5. Chrobak, J., Lorincz, A., & Buzsaki., G. 2000. *Hippocampus*, **10**, 457.
6. Clewley, R., Rotstein, H., & Kopell., N. 2005. *Multiscale Modeling and Simulation*, **4**, 732.
7. Crook, S., Ermentrout, G. B., & Bower, J.M. 1998. *Neural Computation*, **10**, 837.

8. Dayan, P., & Abbott, F. 2005. *Theoretical Neuroscience: Computational and Mathematical Modeling of Neural Systems.* MIT press.
9. Dickson, C., Magistretti, J, Shalinsky, M.H., Fransén, E., Hasselmo, M.E., & Alonso, A. 2000. *Journal of Neurophysiology*, **83**, 2562.
10. Ermentrout, B., Pascal, M., & Gutkin., B. 2001. *Neural Computation*, **13**, 1285.
11. Gloveli, T., Dugladze, T., Rotstein, H., Traub, R., Heinemann, U., Monyer, H., Whittington, M., & Kopell, N. 2005. *Proc. Nat. Acad. Sci.*, **102**, 13295.
12. Izhikevich, E. 2006. *Dynamical Systems in Neuroscience: The Geometry of Excitability and Bursting.* MIT press.
13. Netoff, T., Banks, M., Dorval, A., Acker, C., Hass, J., J., N. Kopell, & White, J. 2005. *Journal of Neurophysiology*, **93**, 1197.
14. Pervouchine, D., Netoff, T., Rotstein, H., Cunningham, M., Whittington, M., White, J., & Kopell, N. 2006. *Neural Computation.*
15. Rotstein, H., Gillies, M., Acker, C., White, J., E.Buhl, Whittington, M., & Kopell, N. 2005. *J. Neurophysiol*, **94**, 1509.
16. Rotstein, H., Oppermann, T., White, J., & Kopell., N. 2006. *J. of Comput. Neurosci.*
17. Saraga, F, Wu, C P, Zhang, L, & Skinner, F K. 2003. *J. Physiol*, **552**, 502.
18. Wang, X. J., & Buzsaki., G. 1996. *J. Neurosci.*, **16**, 6402.
19. White, J., Chow, C., Ritt, J., Soto-Trevino, C., & Kopell, N. 1998. *J. Comput. Neurosci.*, **5**, 5–16.
20. Whittington, M.A., Traub, R.D., Kopell, N., Ermentrout, G.B., & Buhl, E.H. 2000. *Int. J. of Psychophysiology*, **38**, 315–336.

SOME IDEAS ON ACTION POTENTIALS

D. Y. HSIEH
ZHOU PEIYUAN CENTER FOR APPLIED MATHEMATICS,
TSINGHUA UNIVERSITY,
BEIJING,100084, CHINA

It is proposed that the action potentials propagates as ion-acoustic waves in the axons rather than the diffusion process according to the telegraph equation. A non-linear oscillator model is also proposed to account for the spike threshold and the signal coding of the action potentials.

1. INTRODUCTION

The workings of action potentials in neurons can typically sketched as follows[1,2].

At rest a neuron has an excess of positive charges on the outside of the cell membrane and an excess of negative charge on the inside. The charge separation gives rise a difference of electric potential, called the resting membrane potential. Take squid as example, the resting potential is -60mV, the potential outside being set as zero. Sodium and potassium ions play important roles for action potentials. The external sodium ion concentration is 440mM per liter, while the internal concentration is 50mM, resulting a Nernst potential of +55mV. The corresponding numbers for potassium ions are 20mM, 400mM and -75mV respectively. Therefore the distribution of potassium ions is fairly close to equilibrium; while for sodium ions, having a Nernst potential 115mV more positive than the the resting potential, there is large electrochemical gradient to drive them into the cell. Action potential arises when a stimulus causes first the opening of the sodium ion channels resulting an inward Na^+ current and rapid rising of internal potential. The rising of internal potential is arrested when it reaches around 40mV because the closing of the sodium ion channels and the opening of the potassium ion channels. With internal potential now 115mV more positive than the Nernst potential, there will be an outward K^+ current. Then the equilibrium resting potential will be restored. The whole process lasts about 1ms.

The action potential is the conducting signal of the neuron. The conducting signal is *all-or-none*, i.e., stimuli below the threshold do not produce a signal, whereas all stimuli above the threshold produce the same signal. The signals are a series of spikes of the same strength. Once the input signal, which has variable amplitude and duration, surpasses the spike threshold, any further increase in amplitude of the input signal increases the frequency with which the action potential are generated, not their amplitude. There is a transformation from the continuous input to a discrete frequency code at the trigger zone of a sensory neuron. How does it come about?

The action potential with spike heights up to 110 mV, thus making the interior of the axon membrane momentarily positive with respect to outside, has duration only about 1 ms, and can travel down the axon at rate up to 100 meters per second. Again how is it done?

We shall describe the prevailing theories on these questions and offer alternate theories for them in the following.

2. THE TELEGRAPH EQUATION

The prevailing theory on signals travelling down the axons is the cable theory, implying that the axon behaves like a poorly insulated telegraph cable. The derivation of the telegraph equation can be briefly described as follows[2,3,4]:

Let $i(x,t)$ and $V(x,t)$ be the current and voltage at a point x in the axon at any instant t. The fall of voltage in a linear element of length dx at the point x is

$$-dV = iRdx + L\frac{\partial i}{\partial t}dx, \tag{1}$$

where R is the series resistance per unit length and L is the inductance per unit length. Let C be the capacitance per unit length 'to the earth' or across the cell membrane, and G the conductance per unit length, then

$$-di = GVdx + C\frac{\partial V}{\partial t}dx. \tag{2}$$

The relations (1) and (2) are equivalent to the pair of partial differential equations

$$\frac{\partial V}{\partial x} + Ri + L\frac{\partial i}{\partial t} = 0, \tag{3}$$

$$\frac{\partial i}{\partial x} + GV + C\frac{\partial V}{\partial t} = 0, \tag{4}$$

Eliminating i from equations (3) and (4), we obtain

$$\frac{\partial^2 V}{\partial x^2} = LC\frac{\partial^2 V}{\partial t^2} + (RC + LG)\frac{\partial V}{\partial t} + RGV. \tag{5}$$

Equation (5) is known as the telegraph equation.

The inductance L is usually ignored in biophysical literature. Thus the "telegraph equation" becomes

$$\frac{\partial^2 V}{\partial x^2} = RC\frac{\partial V}{\partial t} + RGV. \tag{6}$$

A numerical example from experiments gives the following values of the parameters:$R = 3.06 \times 10^8 \Omega m^{-1}, C = 1.57 \times 10^{-6} F m^{-1}$, and $\frac{1}{G} = 1.27 \times 10^3 \Omega m$.

Equation (6) is essentially a diffusion equation. Let us recall that for diffusion equation

$$D\frac{\partial^2 u}{\partial x^2} = \frac{\partial u}{\partial t}, \tag{7}$$

its Green function is $u_G(x,t) = (4\pi Dt)^{-\frac{1}{2}} \exp(\frac{-r^2}{4Dt})$. The 'diffusion length' which gives a measure of the extent of substantial influence is thus defined as $\sqrt{4Dt}$. And a 'speed' c_D can be estimated as $\sqrt{\frac{4D}{t_c}}$, where t_c is some characteristic time.

Now the diffusion coefficient in equation (6) is $\frac{1}{RC}$. If we use the values of the numerical example, then we find c_D is approximately $\sqrt{\frac{0.01}{t_c}}$. If we take the characteristic time to be 1 millisecond, then the 'speed' c_D is approximately 3 meters per second. It is known that the action potentials 'propagates' at speed between 1 and 100 meters per second. Thus the telegraph equation appears to give a satisfactory explanation of the mechanism of propagation of action potentials.

The word 'propagation' and the description that action potentials travel down the axon without distortion implies that they are waves. Indeed telegraph signals travel down the electric wires as waves, i.e., electromagnetic waves. From the telegraph equation (5), the wave speed c_W is $\sqrt{LC}$. Assume that the axon is nonpermeable, and thus its permeability is 1. Then the inductance L will be $0.5 \times 10^{-7} Hm^{-1}$. With $C = 1.57 \times 10^{-6} Fm^{-1}$, we have c_W to be approximately 3×10^6 meters per second. This value of propagation speed is too high. Therefore diffusion dominates and a diffusive mechanism for transport of action potentials has usually been adopted.

Since the action potentials do propagate like a wave, we suggest an alternative proposal for the mechanism, i.e., they propagate like ion-acoustic waves in plasma.

3. THE ION-ACOUSTIC WAVES

We give a brief presentation of the derivation of the equations governing the propagation of ion-acoustic waves in plasma in Appendix A. We shall see how we can adapt those information to the propagation of action potentials in axons.

In neurons and axons, there are Na^+ and K^+ ions and also negative ions. There are more negative ions than positive ions to give the equilibrium resting potential of -70mV. For the purpose of presenting the essential idea of our theory, let us assume that the equilibrium resting potential is zero. We can consider both the positive and negative ions are at rest at equilibrium. The tendency to maintain this equilibrium will result in, just like the tendency to keep charge neutrality in plasma, $n_i = n_e = n$ and $\mathbf{v}_i = \mathbf{v}_e = \mathbf{v}$. Here the subscript e refers to negative ions rather than electrons. Therefore we again obtain the phase velocity of the ion-acoustic waves (A8):

$$c_p = \sqrt{\frac{\gamma k_B (T_i + T_e)}{m_i + m_e}}. \tag{8}$$

What should be the proper values of γ and $T_{i,e}$? The ions in the axon are not ideal gas. We shall use equation (8) just to have a ball-park estimate. Since mainly the positive ions Na^+ and K^+ are flowing into and out of the neuron, we assume that the negative ions Cl^- do not move much. Thus we shall set $T_e = 0$. Assume that the "gas" behaves isothermally. Then we have $\gamma = 1$. At room temperature, we have $T_i = 300$. Take the molecular weight of the positive ion to be 30 and that of the negative ion 20. then we obtain approximately $c_p = 220$ meter per second, which is in same order of magnitude as the observed value of the propagation speed of the action potential.

We have made use of the well-established analogous presentation of the ion-acoustic waves in plasma to demonstrate plausibility of our suggestion that the mechanism of the propagation of action potential is similar. A more refined analysis and experimental verification is required to establish its validity. Theoretically, we need to deal separately with the momentum and continuity equations of both positive and negative ions as well as the neutral background particles. We also need to deal with the equation governing the electric field E, say, the Gauss law. Much work remains to be done.

4. THE SENSORY CODING

The prevailing theory to explain the transformation from generator potential to frequency code is as follows. A sensory nerve terminal is functionally divided into two regions. There is a receptor region, which is particularly sensitive to the stimulus and responds to it by means of a graded recptor/generator potential, and there is a conductile region whose activity consists of all-or-nothing action potentials. The part of the conductile region next to the receptor region is the impulse initiation site. The all-or-nothing action potentials are successively initiated at the impulse initiation site. Somehow, if the intensity of the generator current is increased, the depolarization process between the successive spike potentials will occur more rapidly.[2].

This theory does not really explain how the generator potential is transformed to the frequency code. We shall now try to propose a model which can account for not only the frequency code but also the spike threshold.

To begin with, take the equation of the linear harmonic oscillator:

$$\frac{d^2u}{dt^2} + \omega^2 u = 0. \tag{9}$$

The frequency of the oscillator will change if ω changes. The larger is the ω, the larger will be the frequency. This is a transformation from amplitude input to frequency code. The model we propose is inspired from this simple idea.

In our model, the open-and-shut of ion channels are considered to behave like a nonlinear oscillator. When there is no stimulus, the channels are executing small amplitude oscillation with some natural frequency. The amplitude is so small that hardly any ions pass through the channels. There is some instability mechanism in the form of 'negative nonlinear spring constant' in the system. This instability mechanism is controlled by the strength of the stimulus. There is also a very stiff built-in nonlinear spring to prevent the run-away of the instability. When the stimulus is small, the small amplitude oscillations of the channels are essentially not affected. However when the stimulus reaches some threshold level, the instability mechanism will cause the channels to open wide, and allow the massive flow of ions to cross the channels, and produce a spike of action potential. The channels are oscillating with large amplitudes and certain frequencies. The frequency of the open-and-shut of the channels depends on the strength of the stimulus. The stronger is the stimulus, the higher is the frequency.

Let us illustrate the general idea by a representative example. Consider a nonlinear oscillator given by the equation:

$$\frac{d^2u}{dt^2} + u - 2b^4u^3 + 16u^{31} = 0, \tag{10}$$

where the second term is responsible for the small amplitude natural oscillation, b in the third term represents the strength of the stimulus causing the instability, and the last term represents the stiff nonlinear spring. Following the general discussions in Appendix B, we find for this case:

$$F(u) = 0.01 - u^2 + b^4u^4 - u^{32}, \tag{11}$$

where we have assigned $c = 0.01$, a small value, to represent the small amplitude natural oscillation. Then for various values of b, we have the following results:

$b = 0.0$	$u_m = 0.1000$	$T = 5.717$
$b = 1.0$	$u_m = 0.1005$	$T = 5.925$
$b = 2.0$	$u_m = 0.1118$	$T = 6.920$
$b = 2.2$	$u_m = 0.1264$	$T = 9.186$
$b = 2.3$	$u_m = 1.1252$	$T = 16.943$
$b = 3.0$	$u_m = 1.1696$	$T = 8.945$

$b = 4.0$	$u_m = 1.2189$	$T = 6.173$
$b = 5.0$	$u_m = 1.2585$	$T = 4.808$
$b = 6.0$	$u_m = 1.2917$	$T = 3.959$
$b = 8.0$	$u_m = 1.3459$	$T = 2.941$
$b = 9.0$	$u_m = 1.3687$	$T = 2.608$

As we may see from the above, for this example, the spike threshold is around $b = 2.25$. For b below this threshold, the opening u_m is around 0.1. For b above this threshold, the opening increase by 10-fold. As b increases further, the period T decreases steadily, or the frequency increases steadily.

This is just a representative example to illustrate the essential features of the basic model. The model can of course be refined to deal with specific real problems.

5. DISCUSSIONS

We have proposed that the action potentials propagate as ion-acoustic waves instead of electromagnetic waves or the diffusion process governed by the telegraph equation. The speed of propagation of electromagnetic waves is much higher than the observed values. On the other hand, if we ignore the magnetic induction, then the mechanism is essentially a diffusion process which does not support wave propagation. Diffusion process represents the transport of physical properties of random microscopic states by molecular collision. If the microscopic states are organized, then the transport by molecular collision will be manifested as acoustic waves. Here the ion motions are organized by the electric field. Thus we have the ion-acoustic wave..

Experimental verifications are still needed to establish the basic soundness of the mechanism of ion-acoustic waves. If the theory turns out to be essentially valid, then more refined analyses could be carried out to deal with the complex real situations. Moreover, the validity of the theory will also imply that mechanical means, other than electric means, play an important role for transmission of neural signals. It also raises the question that whether there are transmission paths other than axons to transmit mechanical signals.

We have also presented a nonlinear oscillator model to account for the spike threshold and the frequency coding of the action potentials. We have suggested that the open-and-shut of ion channels behave like a nonlinear oscillator. This again requires experimental verification. The important point is that such a mechanism can be theoretically constructed. Other things inside the neuron could play the roles of the ion channels in the theory. Again theoretical analysis could be refined to deal with specific real problems.

Appendix A. Ion-Acoustic Waves in Plasm

Consider a plasma consists of electrons and a single species of positive ions. Let $m_{i,e}$, $n_{i,e}$,and $\mathbf{v_{i,e}}$ be the mass, number density, and velocity of the ions and electrons. Then the momentum equations of the ions and electrons are

$$m_{i,e}n_{i,e}\left[\frac{\partial \mathbf{v_{i,e}}}{\partial t}+(\mathbf{v_{i,e}}\cdot\nabla)\mathbf{v_{i,e}}\right]=-\nabla p_{i,e}\pm en_{i,e}\mathbf{E}, \tag{A1}$$

where $p_{i,e}$ is the pressure of the ions and electrons, e the electronic charge, and $\mathbf{E}$ is the electric field which is caused by the charge separation due to the motion of the ions and electrons.

In equilibrium the plasma is charge neutral and thus $n_i = n_e = n$. For plasma, there are two distinct features which enable us to simplify the analysis. One is the slight departure from the charge neutrality. The other is that the electron mass is much smaller than the mass of ion. The approximate charge neutrality gives not only $n_i = n_e = n$, but also $\mathbf{v}_i = \mathbf{v}_e = \mathbf{v}$. Thus we obtain from (A1)

$$\frac{1}{m_i}(-\nabla p_i + en\mathbf{E}) = \frac{1}{m_e}(-\nabla p_e - en\mathbf{E}). \tag{A2}$$

Or

$$\mathbf{E} = \frac{m_e\nabla p_i - m_i\nabla p_e)}{ne(m_i+m_e)}. \tag{A3}$$

From (A1), we can also obtain

$$(m_i+m_e)n\left[\frac{\partial \mathbf{v}}{\partial t}+(\mathbf{v}\cdot\nabla)\mathbf{v}\right]=-\nabla p_i-\nabla p_e. \tag{A4}$$

If we assume that the ions and electrons obey adiabatic ideal gas law, then

$$\nabla p_{i,e} = \gamma k_B T_{i,e} \nabla n, \tag{A5}$$

where $T_{i,e}$ are the ion and electron temperatures,k_B is the Boltzmann constant, and γ is the ratio of specific heats. Inserting (A5) in (A4), we obtain the momentum equation

$$(m_i + m_e)n \left[\frac{\partial \mathbf{v}}{\partial t} + (\mathbf{v} \cdot \nabla)\mathbf{v} \right] = -\gamma k_B (T_i + T_e) \nabla n. \tag{A6}$$

Now we also have the continuity equation

$$\frac{\partial n}{\partial t} + \nabla \cdot (n\mathbf{v}) = \mathbf{0}. \tag{A7}$$

Equations (A6) and (A7) are just the dynamic equations governing an adiabatic fluid. The distinct feature of this formulation is the equation (A3), which relates the electric field to the motion of ions and electrons.

Linearizing the equations (A6) and (A7), and consider plane waves propagating so that n and $\mathbf{v}$ are proportional to $exp[i(\omega t - \mathbf{k} \cdot \mathbf{r})]$. Then we obtain readily that the phase velocity of the ion-acoustic wave is

$$c_p = \frac{\omega}{k} = \sqrt{\frac{\gamma k_B (T_i + T_e)}{m_i + m_e}}. \tag{A8}$$

Appendix B. Non-linear Oscillators

Consider the differential equation:

$$\frac{d^2 u}{dt^2} + f(u) = 0, \tag{B1}$$

where $f(u)$ is real, $f(u) \approx u$, as $u \to 0$, and $f(u)$ is positive as $u \to \infty$.

For linear harmonic oscillators, $f(u) = u$. Multiply (B1) by $\frac{du}{dt}$ and integrate, we obtain

$$\left(\frac{du}{dt}\right)^2 + 2\int_0^u f(x)dx = c, \qquad (B2)$$

where c is the value of $\left(\frac{du}{dt}\right)^2$ when $u = 0$.

Denote

$$F(u) = c - 2\int_0^u f(x)dx, \qquad (B3)$$

then we obtain

$$\frac{du}{dt} = \sqrt{F(u)}. \qquad (B4)$$

Let u_m be the first positive zero of $F(u)$, Then equation (B4) represents oscillations around $u = 0$, between $-u_m$ and u_m. The period of the oscillation T is given by

$$T = 4\int_0^{u_m} \frac{du}{\sqrt{F(u)}}. \qquad (B5)$$

References

1. Eric R., Schwartz, James H. & Jessell, Thomas M., *Essentials of Neural Science and Behavior,* Appleton & Lange , Norwalk, Connecticut (1995).
2. David J.,*The Physiology of Excitable Cells,* Cambridge University Press, Cambridge, UK (1998).
3. Sneddon, Ian N.,*Elements of Partial Differential Equations,* McGraw Hill, New York (1957).
4. Peter & Abbott, L. F., *Theoretical Neural Science: Computational and Mathematical Modelling of Neural System,* MIT Press (2001).

NEGATIVE FEEDBACK IN MORPHOGEN GRADIENTS

M. KHONG AND F.Y.M. WAN

Department of Mathematics and University of California, Irvine, CA 92697-3875.

In this paper, the effects of a conventional form of negative feedback on receptor synthesis rate as a mechanism to induce robustness of biological development of the Drosophila wing imaginal disc is analyzed through the initial-boundary value problem in a basic partial differential equation model for the extracellular biological development activities. The existence, uniqueness, linear stability and monotonicity of the steady state signaling gradient are established rigorously. Solutions are then obtained for a special case of the steady state problem to show explicitly the effects of the chosen form of negative feedback. It is evident from the results that the principal effect of such a feedback mechanism is to render the signaling bound morphogen gradient more uniformly distributed except for a narrow layer adjacent to the edge of the posterior compartment of the wing disc. While the change in the magnitude of the maximum signaling morphogen concentration near the ligand source may be kept at an acceptable level, the leveling and flattening of the gradient render it less differential in space (except in the boundary layer) and hence more deviating from the desired gradient for the target biological pattern. The conclusion is then shown to apply to the general case with the help of the theoretical results on monotonicity. These results suggest that negative feedback on the receptor synthesis rate of the chosen form is not effective for promoting robustness. In fact, it drives the system somewhat in the opposite direction, a conclusion supported by the simulation results. The findings in turn suggest more useful forms of the negative feedback for mediating receptor synthesis to offset a higher ligand synthesis rate for our types of systems.

1. Introduction

Morphogens (also known as *ligands*) are molecular substances (proteins) that bind to selected signaling cell surface *receptors* (and other kinds of non-signaling molecules not considered herein). The concentration gradients of morphogens bound to their associated signaling cell receptors are known to be responsible for differential cell expressions and the patterning of biological tissues during the developmental phase of the biological host. For a number of morphogen families (including *Dpp* in the wing imaginal disc of Drosophila fruit flies), it is well established that a signaling morphogen is produced at a localized source at some synthesis rate V_L and transported away from the source by diffusion (and possibly other transport mechanisms not considered explicitly herein). Some of the transported ligand molecules *bind* with cell receptors along the way forming signaling *ligand-receptor complexes* (called *bound morphogens* or *signaling morphogens* for brevity). Some of the bound morphogens *endocytose* into the cell interior while others *dissociate* to free up ligands to be transported further downstream for possible binding with receptors at other locations. The bound morphogens in the cell interior may *degrade* and dissolve or *exocytose* back to the extracellular space. At any instance in time, the concentration of bound morphogen complexes generally decreases with distance away from the morphogen source and this concentration gradient triggers differential cell expressions resulting in a cell tissue pattern (see [2], [3], [5], [16] and other references cited in [8]). The time evolution of the basic morphogen activities (diffusion, reversible binding with renewable receptors and degradation) in Drosophila wing imaginal discs toward a relatively steady signaling morphogen gradient for cell expression has been investigated recently by systems of partial differential equations and auxiliary conditions that mathematically characterize these basic activities [7,9,17]. By analyzing mathematically the initial-boundary value problems for these models, the effects of various system rate parameters were delineated. In particular, the concentration of bound morphogen complexes was shown to tend respectively to a unique, linearly stable steady state gradient that is monotone decreasing from the localized source to the edge of the imaginal disc (see [7,9,17]).

While the mathematical models of [7,9,17] as well as those of [10] established the consistency of diffusion as a morphogen transport mechanism with experimental observations on signaling morphogen gradients and cell expression, they do not exhibit the expected robustness of biological development with respect to system parameter changes. For example, doubling the ligand synthesis rate (due to a substantial temperature change, say) was found to result in a substantial change in model response in the magnitude and shape of the concentration gradients [12]. A numerical measure E was introduced in [12] for characterizing robustness, with $E = 0$ being perfectly robust and $E \leq 0.2$ taken (somewhat arbitrarily) to be acceptably robustness. Numerically simulated responses to 10^6 different sets of parameter values in the six dimensional parameter space are seen to be non-robust with $E \geq ln(2)/ln(5) \gtrsim 0.43$ in all cases. Evidently, additional biological processes are at work and must be included in the model to ensure robustness of the development with respect to changes in system properties.

In [12], negative feedback, an often used mechanism for mediating excessive changes, was applied to the receptor synthesis rate V_R in the form

$$V_R = V_{\min} + \frac{V_{\max} - V_{\min}}{1 + \Gamma[LR]^n} \tag{1}$$

where $V_{\max}$ and $V_{\min}(< V_{\max})$ are the maximum and minimum synthesis rates for receptors, $[LR]$ is the total concentration of signaling (bound) morphogens (both in the cell interior and the extracellular space), and Γ and n are two constant parameters with the latter generally referred to as the Hill's coefficient [4]. Evidently, we have $V_R = V_{\max}$ in the absence of bound morphogens and V_R tends to $V_{\min}$ as $[LR]$ tends to infinity. It was expected that at high morphogen synthesis rate giving rise to a high transient concentration of bound morphogens, the feedback mechanism (1) would reduce the receptor synthesis rate to a substantially lower level to result in concentration gradients differing insignificantly from the response to the normal synthesis rate prior to the rate change and thereby ensuring robustness. Rather surprisingly, the results of numerical simulations for 10^6 sets of parameter values for a system with feedback mechanism (1) show no improvement in the robustness in the model response to a doubling of ligand synthesis rate. In fact, more parameter value sets with larger E values are found in the simulation results for the model with negative feedback.

In this paper, we analyze the effects of a negative feedback of the form (1) on the response of the model system. We will establish that similar to the original system investigated in [9], the initial-boundary value problem for the model system with feedback is well-posed. More specifically, we will prove the existence, uniqueness and linear stability of the monotone decreasing steady state signaling gradient. The various proofs for the present system are more intricate than those in [9] for reasons that will become apparent after we have formulated the mathematical problem. We then obtain useful solutions of the problem for the special case of $n = 1$ which will provide us insight to the effects of our particular type of negative feedback. It will be seen from the results that the principal effect of such a feedback mechanism is to render the signaling bound morphogen gradient $[LR]$ more uniformly distributed except for a boundary layer adjacent to the edge of the wing imaginal disc. While the change in the magnitude of the maximum bound morphogen concentration near the ligand source may be kept at an acceptable level by our negative feedback, the leveling and flattening of the gradient render the tissue patterning less differential in space and hence deviate more from the target biological patterning.

The conclusions drawn from the solution for the $n = 1$ case will then be extended to the original model system with a general Hill's coefficient ($n \geq 1$). This suggests that negative feedback on the receptor synthesis rate of the form (1) is not effective for inducing robustness. In fact, it drives the system somewhat in the opposite direction, a conclusion supported by the simulation results of [12]. The findings in turn suggest more appropriate forms of the negative feedback for mediating receptor synthesis should be explored for offsetting a higher ligand synthesis rate. The effects of some of these forms of negative feedback on robustness are being examined.

2. An Extracellular Formulation with Receptor Synthesis

As in [8,9], we simplify the development of the wing imaginal disc of a Drosophila fly as a one-dimensional phenomenon. In doing so, we ignore variations in the ventral-dorsal direction and the apical-basal direction since extensions of the one-dimensional model to account for developments in these other directions are straightforward (see [11,17,18] for example). To investigate the consequences of negative feedback of signaling morphogen concentration on the receptor synthesis rate, we will work with an extracellular formulation similar to [9]. As we have shown in [10,13], the results for such a model may be re-interpreted as the corresponding results for a model where morphogen-receptor complexes internalize (through endocytosis) before degradation.

To simplify our discussion, we note that the morphogen production zone divides the wing imaginal disc into the anterior compartment and the posterior compartment. We consider in this paper the part of the wing disc extending from the midpoint, $X = -X_m$, of the *Dpp* production zone to the edge of the posterior compartment at $X = X_{\max}$ with morphogen produced only in $-X_{\min} < X < 0$. Let $[L(X,T)]$ be the concentration (in micromoles) of the diffusing morphogen *Dpp* at time T and location X. Let $[R(X,T)]$ and $[LR(X,T)]$ be the concentration of unoccupied receptors and morphogen occupied receptors (or bound morphogens), respectively. For the underlying biological processes of the development described in [8,9], we add to Fick's second law for diffusive transport of *Dpp* ($\partial[L]/\partial T = D\partial^2[L]/\partial X^2$, D being the diffusion coefficient) terms that incorporate the rate of morphogen binding with receptors, $-k_{on}[L][R]$, and dissociation, $k_{off}[LR]$, with k_{on} and k_{off} being the binding rate constant and dissociation rate constant, respectively. In living tissues, molecules that bind receptors do not simply stay bound, some will dissociate and others will (endocytose and) degrade [16]. In accounting for the time rate of change of the *Dpp*-receptor complexes, we allow for constitutive degradation of $[LR]$ by introducing a degradation rate term with a rate constant k_{deg}. There is also a separate accounting of the time rate of change of the concentration of unoccupied receptors as they are being synthesized and degrade continuously in time (with a degradation rate constant r_{deg} as in [9]). In this way, we obtain the following reaction-diffusion system for the evolution of three concentrations $[L]$, $[LR]$ and $[R]$:

$$\frac{\partial[L]}{\partial T} = D\frac{\partial^2[L]}{\partial^2 X} - k_{on}[L][R] + k_{off}[LR] + V_L(X,T) \tag{2}$$

$$\frac{\partial[LR]}{\partial T} = k_{on}[L][R] - (k_{off} + k_{\text{deg}})[LR] \tag{3}$$

$$\frac{\partial[R]}{\partial T} = V_R(X,T) - k_{on}[L][R] + k_{off}[LR] - r_{\text{deg}}[R] \tag{4}$$

for $-X_{\min} < X < X_{\max}$ and $T > 0$ where $V_L(X,T)$ and $V_R(X,T)$ are the rate at which *Dpp* and receptors are synthesized, respectively.

In [7,8,10], we were interested only in the portion of the wing disc corresponding to $X > 0$ where there is no morphogen production (so that $V_L(X,T) = 0$ for $X > 0$) with the introduction of *Dpp* into the region $0 < X < X_{\max}$ characterized by a point source at the end $X = 0$. A model with a finite *Dpp* synthesis region of the form (2) - (4) but without feedback has been investigated in [9] where the relation between that model and point source models was discussed. Here, we add to the finite *Dpp* production region model of [9] negative feedback of the $[LR]$ concentration on the receptor synthesis rate in the form (1) where the Hill's coefficient n and multiplier Γ are constants to be specified.

In the absence of morphogens so that $[LR] = 0$, we take $V_R(X,T) = V_{\max}(X)$ in order to have a steady state receptor concentration for that case. In the limit as $[LR] \longrightarrow \infty$, we require that V_R tends to $V_{\min}(X)$ ($< V_{\max}(X)$) to allow for the possibility of steady state $[L]$, $[R]$, and $[LR]$ concentrations also.

With $-X_{\min}$ being the midpoint of the *Dpp* production region, we have by symmetry

$$X = -X_{\min}: \ \frac{\partial [L]}{\partial X} = 0 \qquad (T > 0). \tag{5}$$

The far end of the wing disc, i.e., the edge of the posterior chamber, is taken to be a sink so that

$$X = X_{\max}: \ [L] = 0 \qquad (T > 0). \tag{6}$$

At $T = 0$, we have the initial conditions

$$[L] = [lR] = 0, \qquad [R] = R_i(X) \qquad (-X_{\min} < X < X_{\max}) \tag{7}$$

where $R_i(X)$ is some initial distribution of signaling receptors.

To reduce the number of parameters in the problem, we introduce a reference unoccupied receptor concentration level $\bar{R}_0$ (to be specified later) and the normalized quantities

$$t = \frac{D}{X_{\max}^2} T, \qquad x = \frac{X}{X_{\max}}, \qquad x_m = \frac{X_{\min}}{X_{\max}}, \tag{8}$$

$$\{a, b, r, r_i\} = \frac{1}{\bar{R}_0}\{[L], [LR], [R], R_i\}, \qquad \gamma = \bar{R}_0 \Gamma, \tag{9}$$

$$\{v_L, v_R, v_{\min}, v_{\max}, \rho_R\} = \frac{1}{D/X_{\max}^2}\left\{\frac{V_L}{\bar{R}_0}, \frac{V_R}{\bar{R}_0}, \frac{V_{\min}}{\bar{R}_0}, \frac{V_{\max}}{\bar{R}_0}, \frac{V_{\min}}{V_{\max}}\right\}, \tag{10}$$

$$\{f_0, g_0, g_r, h_0\} = \frac{1}{D/X_{\max}^2}\{k_{off}, k_{\deg}, r_{\deg}, k_{on}\bar{R}_0\}. \tag{11}$$

In terms of these scaled quantities, we write the IBVP in the following normalized form

$$\frac{\partial a}{\partial t} = \frac{\partial^2 a}{\partial x^2} - h_0 a r + f_0 b + v_L(x, t) \qquad (-x_m < x < 1) \tag{12}$$

$$\frac{\partial b}{\partial t} = h_0 a r - (f_0 + g_0) b, \qquad \frac{\partial r}{\partial t} = v_R(x, t) - h_0 a r - g_r r + f_0 b \qquad (-x_m \le x \le 1) \tag{13}$$

with

$$x = -x_m: \ \frac{\partial a}{\partial x} = 0, \qquad x = 1: \ a = 0 \tag{14}$$

for $t > 0$ and

$$t = 0: \ a = b = 0, \qquad r = r_i(x) \qquad (-x_m < x < 1) \tag{15}$$

where

$$v_R = \frac{V_{\max}/\bar{R}_0}{D/X_{\max}^2}\left[\rho_R + \frac{1-\rho_R}{1+\gamma b^n}\right]$$
$$= v_{\max}(x,t)\left[\frac{1+\rho_R\gamma b^n}{1+\gamma b^n}\right] \equiv v_{\max}(x,t) f_R(b), \tag{16}$$

and

$$0 \le \rho_R = \frac{V_{\min}(X,T)}{V_{\max}(X,T)} = \frac{v_{\min}(x,t)}{v_{\max}(x,t)} \le 1. \tag{17}$$

3. Time Independent Steady State Behavior

3.1. *Reduction of the Steady State Equations*

For cells to express differentially, it is important that the signaling morphogen concentrations in the wing imaginal disc evolves toward to a time independent steady state with a reasonable slope and convexity. For the present investigation, it suffices to consider a localized *Dpp* synthesis rate in the form of a step function with $V_L(X,T) = \bar{V}_L H(-X)$ for some constant $\bar{V}_L$. Correspondingly, we have

$$v_L(x,t) = v_L(x) = \bar{v}_L H(-x) = \begin{cases} \bar{v}_L \\ 0 \end{cases}, \qquad \bar{v}_L = \frac{\bar{V}_L/\bar{R}_0}{D/X_{\max}^2}. \tag{18}$$

We will also consider uniform maximum and minimum receptor synthesis rates with

$$\{V_{\max}(X,T),\, V_{\min}(X,T)\} = \{\bar{V}_{\max},\, \bar{V}_{\min}\} \tag{19}$$

so that the normalized nonnegative receptor synthesis rate v_R takes the form

$$v_R(x,t) = \frac{\bar{V}_{\max}/\bar{R}_0}{D/X_{\max}^2} f_R(b) \equiv \bar{v}_{\max} f_R(b) \tag{20}$$

where we have now a constant ratio of the two limit receptor synthesis rates: $\rho_R = \bar{V}_{\min}/\bar{V}_{\max} \equiv \bar{v}_{\min}/\bar{v}_{\max}$. Note that the case of different receptor synthesis rates in and outside the morphogen production zone has been examined in previous publications [7,8]. Here we focus only on the case where $V_{\min}$ and $V_{\max}$ are constants so that, in the absence of feedback, the receptor synthesis rate is uniform throughout the posterior compartment (given that we have $f_R(b) = 1$ for $\gamma = 0$).

With the initial receptor concentration taken to be the steady state receptor distribution prior to the onset of morphogen production, $R_i(x) = [V_R(X)/r_{\deg}]_{[LR]=0}$, we set

$$\bar{R}_0 = \frac{\bar{V}_{\max}}{r_{\deg}} \tag{21}$$

so that we have from (16)

$$\bar{v}_{\max} = g_r, \qquad R_i(x) = \bar{R}_0 r_i(x). \tag{22}$$

For our choice of receptor synthesis rate $V_R(X)$, we have

$$r_i(x) = \left[\frac{V_R(X)}{\bar{V}_{\max}}\right] = 1 \tag{23}$$

given that we have taken $V_R(X) = V_{\max}(X) = \bar{V}_{\max}$ in the absence of bound morphogen concentration ($b = 0$).

We are interested in a time independent steady state solution

$$\{a(x,t), b(x,t), r(x,t)\} = \{\bar{a}(x), \bar{b}(x), \bar{r}(x)\} \tag{24}$$

for the system (9) - (12). For such a solution, we may set all time derivatives in these equations to zero to get

$$0 = \bar{a}'' - h_0\bar{a}\bar{r} + f_0\bar{b} + v_L(x)\,, \tag{25}$$

$$0 = h_0\bar{a}\bar{r} - (f_0 + g_0)\bar{b}, \qquad 0 = g_r f_R(\bar{b}) - h_0\bar{a}\bar{r} - g_r\bar{r} + f_0\bar{b}\,, \tag{26}$$

where a prime indicates differentiation with respect to x. The nonlinear second order system of ODE (25) - (26) is augmented by the boundary conditions

$$\bar{a}'(-x_m) = 0, \qquad \bar{a}(1) = 0. \tag{27}$$

With $v_L(x)$ being piecewise constant, the form of (25) - (26) requires that $\bar{a}(x)$ and its first derivative to be continuous at $x = 0$.

In previous investigations without feedback [7,8,9,10], the two equations in (26) were solved for $\bar{b}$ and $\bar{r}$ in terms of $\bar{a}$ and the results used to reduce (25) to a single ODE for $\bar{a}$. For the present problem, we can solve the first equation in (26) to get $\bar{r} = \alpha_0\bar{b}/\bar{a}$. Upon substituting this in the second equation of (26), we obtain a polynomial equation in $\bar{b}$ with $\bar{a}$ in the various coefficients:

$$\begin{aligned} P(\bar{b}) &\equiv \gamma A(\bar{a})\bar{b}^{n+1} - \gamma B(\bar{a})\bar{b}^n + A(\bar{a})\bar{b} - [B(\bar{a}) + C(\bar{a})] \\ &= (A\bar{b} - B)(\gamma\bar{b}^n + 1) - C = 0 \end{aligned} \tag{28}$$

where

$$A(\bar{a}) = g_0h_0\bar{a} + g_r(f_0 + g_0) \equiv A_1\bar{a} + A_0, \tag{29}$$

$$B(\bar{a}) = h_0\bar{v}_{\min}\bar{a} \equiv B_1\bar{a}, \qquad C(\bar{a}) = h_0\bar{a}(\bar{v}_{\max} - \bar{v}_{\min}) \equiv C_1\bar{a}\,. \tag{30}$$

Except for the special case $n = 1$, the relation (28) cannot be used to express $\bar{b}$ explicitly in terms of $\bar{a}$.

On the other hand, $P(\bar{b})$ is linear in $\bar{a}$ so that we can use (28) to express $\bar{a}$ in terms of $\bar{b}$:

$$\bar{a} = \frac{\alpha_0\bar{b}}{f_R(\bar{b}) - \varsigma\bar{b}} \tag{31}$$

where

$$\alpha_0 = \frac{f_0 + g_0}{h_0}, \qquad \varsigma = \frac{g_0}{g_r}. \tag{32}$$

The expression (31) can then be used to eliminate $\bar{a}$ from (25) to get a single second order ODE for $\bar{b}$:

$$\left[\frac{\alpha_0 \bar{b}}{f_R(\bar{b}) - \varsigma \bar{b}}\right]'' - g_0 \bar{b} + v_L(x) = 0 \qquad (-x_m < x < 1), \tag{33}$$

keeping in mind that $f_R(\bar{b})$ depends on x through $\bar{b}(x)$ (see (16)). Unfortunately, the form of this ODE is awkward both for theoretical analysis and numerical solutions even if the boundary conditions for $\bar{b}$ also take relatively simple forms:

$$\bar{b}'(-x_m) = 0, \qquad \bar{b}(1) = 0. \tag{34}$$

3.2. *Existence of a Unique Set of Steady State Gradients*

Whether we work with $\bar{a}(x)$ or $\bar{b}(x)$ as the primary unknown, it is not enough to compute solutions of the BVP governing the steady state gradients for some sets of values for the system rate constants. Biologists want to be assured that such steady state gradients exist for any biologically realistic set of parameter values. For this and other reasons, we will show in this section that there is a unique solution of the relevant BVP for the steady state morphogen concentrations. Since the auxiliary conditions are naturally prescribed in terms of the free morphogen concentration, we will stay with the unknown $\bar{a}(x)$ and take $\bar{b}(x)$ and $\bar{r}(x)$ in terms of $\bar{a}(x)$. For this purpose, we need the following preliminary result:

Lemma 3.1. *For a fixed γ and n, there exists a unique $\bar{b}$ in $(B(\bar{a})/A(\bar{a}), \infty)$ for any nonnegative $\bar{a}$, denoted by $b^*(\bar{a})$, which is an increasing function of $\bar{a}$.*

Proof. Since $\bar{b}$ is non-negative, we have from (28) $\bar{b} = b^*(0) = 0$ and $P(\bar{b}) \leq -C(\bar{a}) \leq 0$ for $\bar{a} > 0$ and $0 < \bar{b} \leq B(\bar{a})/A(\bar{a})$. For larger values of $\bar{b}$, $P(\bar{b})$ is strictly increasing for all $\bar{b}$ in $(B(\bar{a})/A(\bar{a}), \infty)$ with $P(\bar{b}) \to \infty$ as $\bar{b} \to \infty$. Since $P(\bar{b})$ is a polynomial, we have by intermediate value theorem a value b^* in $(B/A, \infty)$, depending on $\bar{a}$, for which $P(\bar{b})$ vanishes, i.e., $P(b^*) = 0$. Furthermore, $\bar{b} = b^*(\bar{a})$ is unique for any $\bar{a}(x) \geq 0$ by the monotonicity of $P(\bar{b})$ in $(B/A, \infty)$. $\square$

Lemma 3.2. *For fixed γ and n, $b^*(\bar{a})$ is an increasing function of $\bar{a}$ for all $\bar{a} > 0$.*

Proof. To show that $\bar{b} = b^*(\bar{a})$ is an increasing function of $\bar{a}$, we differentiate (28) with respect to $\bar{a}$ to get

$$\frac{db^*}{d\bar{a}} = \frac{A_0(1 + \gamma \bar{b}^n)}{\Delta} \frac{\bar{b}}{\bar{a}}, \qquad \Delta = \frac{(1 + \gamma \bar{b}^n)}{(1 + \gamma \bar{b}^n)^2 A + n\gamma \bar{b}^{n-1} C}. \tag{35}$$

Since the right side of the first relation above is positive for $\bar{b}$ in $(B/A, \infty)$, the second part of the lemma is proved. □

With Lemma 3.1 above, we can write

$$-\bar{a}'' + g_0 b^*(\bar{a}) - v_L(x) = 0 \qquad (-x_m < x < 1), \tag{36}$$

$$\bar{a}'(-x_m) = 0, \qquad \bar{a}(1) = 0 \tag{37}$$

where $b^*(\bar{a})$ a well-defined continuous and monotone function of $\bar{a}$. Hence we have a well-defined BVP for $\bar{a}(x)$ for which we will show presently that there is a unique monotone solution.

Proposition 3.1. *The BVP (36)-(37) has a nonnegative solution* $\bar{a}(x) \geq 0$.

Proof. The existence of a nonnegative solution of the boundary value problem is proved by producing a nonnegative upper solution and a nonnegative lower solution for the problem. From (28), we have $P(\bar{b}) \equiv (A\bar{b} - B)(\gamma \bar{b}^n + 1) = C$ so that $\bar{b} \leq (B + C)/A$ and therewith

$$-a_\ell'' + g_0 \frac{B+C}{A} - v_\ell(x) \leq -a_\ell'' + g_0 b^*(a_\ell) - v_\ell(x) \qquad (-x_m < x < 1).$$

Hence, a lower solution a_ℓ of the BVP for $\bar{a}$ is therefore given by the solution of

$$-a_\ell'' + g_0 \frac{B(a_\ell) + C(a_\ell)}{A(a_\ell)} - v_\ell(x) = 0, \qquad a_\ell'(-x_m) = 0, \qquad a_\ell(1) = 0.$$

For an upper solution, we note from $C \geq 0$ and (28) that $b(a) \geq B(a)/A(a)$ so that

$$-a_u'' + g_0 \frac{B(a_u)}{A(a_u)} - v_\ell(x) \geq -a_u'' + g_0 b^*(a_u) - v_\ell(x) \qquad (-x_m < x < 1).$$

An upper solution a_u of the BVP for $\bar{a}$ is therefore given by the solution of the

$$-a_u'' + g_0 \frac{B(a_u)}{A(a_u)} - v_\ell(x) = 0, \qquad a_u'(-x_m) = 0, \qquad a_u(1) = 0.$$

By the results of [9], we know both $a_\ell(x)$ and $a_u(x)$ exist and are nonnegative. A theorem of D.H. Sattinger established in [14] (see also [1,15]) assures the existence of a nonnegative $\bar{a}(x)$ for the BVP (36) - (37) with

$$0 \leq a_\ell(x) \leq \bar{a}(x) \leq a_u(x).$$

□

Proposition 3.2. *The solution of the BVP for* $\bar{a}(x)$ *is unique.*

Proof. Let $a_1(x)$ and $a_2(x)$ be two (nonnegative) solutions and $a(x) = a_1(x) - a_2(x)$. Then as a consequence of the differential equation (33) for $a_1(x)$ and $a_2(x)$, the difference $a(x)$ satisfies the following differential equation:

$$-a'' + g_0[b^*(a_1) - b^*(a_2)] = -a'' + g_0 b^{\cdot}(\xi) a = 0$$

where $b^{\cdot}(a) = db^*/da$ and where we have used the mean value theorem for some intermediate value ξ between a_1 and a_2. Form

$$\int_{-x_m}^{1} [-\, a'' + g_0 b^{\cdot}(\xi) a]\, a(x) dx = 0.$$

Upon integration by parts and applications of the boundary conditions in (37), the relation above may be transformed into

$$\int_{-x_m}^{1} [a']^2 dx + g_0 b^{\cdot}(\xi) \int_{-x_m}^{1} [a(x)]^2 dx = 0.$$

Given Lemma 3.2, the left side of the equation above is nonnegative and vanishes only if $a(x) \equiv 0$ and hence uniqueness. □

3.3. *Monotonicity*

Proposition 3.3. *$\bar{a}(x)$ is monotone decreasing in $(0, 1)$.*

Proof. First, we prove that $\bar{a}(x)$ does not have a local maximum in $(0, 1)$. If an interior maximum occurs at x_0, then $a'(x_0) = 0$ and $a''(x_0) \leq 0$. But we have from (36) $\bar{a}''(x_0) = g_0 b^*(\bar{a}) \geq 0$. Hence, we must have $a''(x_0) = 0$ and therewith $b^*(\bar{a}(x_0)) = 0$. In that case (28) reduces to $B(x_0) + C(x_0) = 0$ or, with the expressions for B and C in (30), $\bar{a}(x_0) = 0$. But $\bar{a}(x_0) = 0$ is a local maximum and $\bar{a}(x) \geq 0$, it follows that $\bar{a}(x) = 0$ for all x in $(0, 1)$. But continuity conditions on $\bar{a}$ and $\bar{a}'$ at $x = 0$ determines $\bar{a}(x)$ for $(-x_m, 0)$, contradicting the ODE for $\bar{a}(x)$ in that interval where $v_L(x)$ does not vanish identically.

On the other hand, $\bar{a}(x)$ does not have a local minimum in $(0, 1)$. If it should attain a minimum at x_1, then $\bar{a}(x)$ must attain a maximum at a point x_2 in $(x_1, 1)$ or $\bar{a}(x) = 0$ in $(0, 1)$, given $\bar{a}(x) \geq 0$ and $\bar{a}(1) = 0$. Neither is possible. Thus, $\bar{a}(x)$ must be monotone in $(0, 1)$. Since $\bar{a}(1) = 0$, $\bar{a}(x)$ must be monotone decreasing. □

Proposition 3.4. *$\bar{a}(x)$ is monotone in $(-x_m, 0)$.*

Proof. First, $\bar{a}(x)$ does not have a local maximum in $(-x_m, 0)$. If there should be at least one interior extremum, consider the one closest to $-x_m$, say at $-x_0$ with $\bar{a}'(-x_0) = 0$. Then the ODE together with the two auxiliary conditions $\bar{a}'(-x_0) = 0$ and $\bar{a}(1) = 0$, completely determines $\bar{a}(x)$ (uniquely by the result of the previous subsection) in $(-x_0\,, 1)$. By continuity, the terminal values $\bar{a}(-x_0)$ and $\bar{a}'(-x_0)$ determines $\bar{a}(x)$ uniquely in the interval at $[-x_m, -x_0]$.

If $\bar{a}(-x_0)$ should be a local maximum then the ODE in this interval requires $\bar{a}(-x_m)$ to be a local maximum also. In that case, there must be a local minimum inside the interval $(-x_m, -x_0)$ which contradicts the stipulation that fact that $\bar{a}(-x_0)$ is closest extremum to the end point $x = -x_m$.

If $\bar{a}(-x_0)$ should be a local minimum, then either $\bar{a}(x)$ attains a local maximum at some x_1 in $(-x_0, 1)$ or $\bar{a}(x) \equiv 0$ there. Neither is possible. The former is impossible given

$$\bar{a}''(-x_1) = g_0 b^*(\bar{a}(-x_1)) - \bar{v}_L \geq g_0 b^*(\bar{a}(-x_0)) - \bar{v}_L = \bar{a}''(-x_0) \geq 0$$

with $\bar{a}(x_1) \geq \bar{a}(x_0)$. The latter is impossible because $v_L(x) > 0$ in $(-x_m, 0)$. Thus, $\bar{a}(x)$ must be monotone in $(-x_m, 0)$. □

Proposition 3.5. *$\bar{a}(x)$ is monotone decreasing in $(-x_m, 1)$.*

Proof. Since $\bar{a}'(0) < 0$ by Proposition 3.3 and $\bar{a}'(x)$ is continuous at $x = 0$, we must have $\bar{a}'(x) < 0$ in $(-x_m, 0]$ given that $\bar{a}(x)$ has no interior minimum or maximum there. It follows that $\bar{a}(x)$ is monotone decreasing in $(-x_m, 0)$ and, by Proposition 3.3, in the larger interval $(-x_m, 1)$. □

4. Linear Stability

4.1. *A Nonlinear Eigenvalue Problem*

In addition to the existence of unique steady state concentrations $\bar{a}(x), \bar{b}(x)$,and $\bar{r}(x)$, it is important for these concentrations to be asymptotically stable. To investigate the stability of the steady state solution known to exist from Theorem 3.1, we consider small perturbations from the steady state solution in the form

$$\{a(x,t), b(x,t), r(x,t)\} = \{\bar{a}(x), \bar{b}(x), \bar{r}(x)\} + e^{-\lambda t}\{\hat{a}(x), \hat{b}(x), \hat{r}(x)\}\,. \tag{38}$$

After linearization, the differential equations (12)-(13) become

$$-\lambda\hat{a} = \hat{a}'' - h_0(\bar{r}\hat{a} + \bar{a}\hat{r}) + f_0\hat{b}\,, \tag{39}$$

$$-\lambda\hat{b} = h_0(\bar{r}\hat{a} + \bar{a}\hat{r}) - (f_0 + g_0)\hat{b}\,, \tag{40}$$

$$-\lambda\hat{r} = -h_0(\bar{r}\hat{a} + \bar{a}\hat{r}) - g_r\hat{r} + [f_0 - p(\bar{b})]\hat{b}\,, \tag{41}$$

where

$$p(b) = n\gamma(\bar{v}_{\max} - \bar{v}_{\min})\frac{\bar{b}^{n-1}}{(1+\gamma\bar{b}^n)^2} \geq 0. \tag{42}$$

The relations (40) and (41) are then solved for $\hat{b}$ and $\hat{r}$ in terms of $\hat{a}$ to get

$$\hat{r} = \frac{h_0\bar{r}(x)[\lambda - p(\bar{b}) - g_0]\hat{a}}{(g_r - \lambda)(f_0 + g_0 - \lambda) + h_0\bar{a}(x)[g_0 - \lambda + p(\bar{b})]}, \tag{43}$$

$$\hat{b} = \frac{h_0\bar{r}(x)(g_r - \lambda)\hat{a}}{(g_r - \lambda)(f_0 + g_0 - \lambda) + h_0\bar{a}(x)[g_0 - \lambda + p(\bar{b})]}. \tag{44}$$

The expressions (43) and (44) are used to eliminate $\hat{b}$ and $\hat{r}$ from (39) to obtain

$$\hat{a}'' + [\lambda - q_r(x;\lambda)]\,\hat{a} = 0, \tag{45}$$

where

$$q_r(x;\lambda) = \frac{h_0\bar{r}(x)(g_r-\lambda)(g_0-\lambda)}{(g_r-\lambda)(g_0+f_0-\lambda)+h_0\bar{a}(x)[g_0-\lambda+p(\bar{b})]}.$$

The ODE for $\hat{a}(x)$ is supplemented by the boundary conditions

$$\hat{a}'(-x_m) = 0, \qquad \hat{a}(1) = 0. \tag{46}$$

Together, (45) and (46) define an eigenvalue problem with λ as the eigenvalue parameter. Though the ODE is linear, the eigenvalue problem is nonlinear since λ appears nonlinearly in $q_r(x;\lambda)$ so that (45) and (46) is not a Sturm-Liouville problem. In the next subsection, we will show that the eigenvalues of the homogeneous boundary value problem defined by the differential equation (45) and the homogeneous boundary conditions (46) must be positive. It follows then that the steady state gradients are asymptotically stable according to linear stability theory.

4.2. *Positive Eigenvalues and Asymptotic Stability*

We will prove linear stability of the steady state solution in two steps. First, we prove that the eigenvalues of (45) and (46) are real. These real eigenvalues are then proved to be positive.

Lemma 4.1. *All the eigenvalues of the nonlinear eigenvalue problem (45) and (46) are real.*

Proof. Suppose λ is a complex eigenvalue and $a_\lambda(x)$ an associated nontrivial eigenfunction, then λ^* is also an eigenvalue with eigenfunction $a_\lambda^*(x)$ where $(\ )^*$ is the complex conjugate of $(\ )$. The bilinear relation

$$\int_{-x_m}^{1} [(a_\lambda^*)a_\lambda'' - (a_\lambda^*)''a_\lambda]dx = 0$$

(which can be established by integration by parts and applications of the boundary conditions in (46)) requires

$$\int_{-x_m}^{1} \{(\lambda-\lambda^*) - [q_r(x;\lambda) - q_r(x;\lambda^*)]\}(a_\lambda^* a_\lambda)dx = 0. \tag{47}$$

It is straightforward to verify that $q_r(x;\lambda) - q_r(x;\lambda^*) = -(\lambda-\lambda^*)\Phi(x;\lambda\lambda^*)$, where

$$\Phi(x;\lambda) = \frac{f_0\,|g_r-\lambda|^2 + h_0\bar{a}(x)\,|g_0-\lambda|^2 + h_0\bar{a}(x)p(\bar{b})[g_0+g_r+2\mathrm{Re}(\lambda)]}{\left|(g_r-\lambda)(g_0+f_0-\lambda)+h_0\bar{a}(x)[g_0-\lambda+p(\bar{b})]\right|^2}, \tag{48}$$

is a positive real value function of λ, given the definition of $p(x)$ in (42). In that case, the condition (47) becomes

$$(\lambda-\lambda^*)\int_{-x_m}^{1} a_\lambda a_\lambda^*[1+\Phi(x;\lambda)]dx = 0. \tag{49}$$

Since the integral is positive for any nontrivial $a_\lambda(x)$, we must have $\lambda - \lambda^* = 0$. Hence, λ does not have an imaginary part. □

Theorem 4.1. *All eigenvalues of the nonlinear eigenvalue problem (39)-(41) and (46) are positive and the steady state concentrations $\bar{a}(x), \bar{b}(x)$ and $\bar{r}(x)$ are asymptotically stable with respect to small perturbations from the steady state.*

Proof. Suppose $\lambda \leq 0$. Let $\hat{a}_\lambda(x)$ be a nontrivial eigenfunction of the homogeneous BVP (45) and (46) for the nonpositive eigenvalue λ. Multiply (45) by $\hat{a}_\lambda$ and integrate over the solution domain to get

$$\int_{-x_m}^{1} \left\{\hat{a}_\lambda \hat{a}_\lambda'' - q_r(x;\lambda)(\hat{a}_\lambda)^2\right\} dx = -\lambda \int_{-x_m}^{1} (\hat{a}_\lambda)^2 dx \,.$$

After integration by parts and applications of the homogeneous boundary conditions (46), we obtain

$$\lambda \int_{-x_m}^{1} (\hat{a}_\lambda)^2 dx = \int_{-x_m}^{1} (\hat{a}_\lambda')^2 dx + \int_{-x_m}^{1} q_r(x;\lambda)(\hat{a}_\lambda)^2 dx. \tag{50}$$

With $\lambda = -|\lambda| \leq 0$, we have

$$q_r(x; -|\lambda|) = \frac{\bar{r}(x) h_0 (g_0 + |\lambda|)(g_r + |\lambda|)}{(g_r + |\lambda|)(g_0 + f_0 + |\lambda|) + h_0 \bar{a}(x)[g_0 + |\lambda| + p(\bar{b})]} \geq 0$$

given the definition of $p(x)$ in (42). For any nontrivial solution of the eigenvalue problem under the assumption $\lambda \leq 0$, the right-hand side of (50) is positive which contradicts the assumption $\lambda = -|\lambda| \leq 0$. Hence the eigenvalues of the eigenvalue problem (45) and (46) must be positive and the theorem is proved. □

4.3. *A Rayleigh Quotient*

Similar to the case of no feedback in [9], we want to know the actual magnitude of the *smallest* eigenvalue to give some idea of how quickly the system returns to steady state after small perturbations. As parametric studies require that we repeatedly compute the time evolution of the concentration of both free and bound morphogens from their initial conditions, the value of the smallest eigenvalue will also give some idea of the decay rate of the transient behavior and thereby the time to reach steady state.

Let λ (> 0) be an eigenvalue of (45) and (46) and $\hat{a}_\lambda(x)$ the corresponding eigenfunction. Upon multiplying ODE (45) for this eigen-pair by $\hat{a}_\lambda(x)$ and integrating by parts, we obtain the following Rayleigh quotient-like relation for λ after observing the boundary conditions (46) which apply to $\hat{a}_\lambda(x)$:

$$\lambda \int_{-x_m}^{1} (\hat{a}_\lambda)^2 dx = \int_{-x_m}^{1} (\hat{a}_\lambda')^2 dx + \int_{-x_m}^{1} q_r(x;\lambda)(\hat{a}_\lambda)^2 dx \,. \tag{51}$$

The following key result can be proved similar to that in [9]:

Lemma 4.2. *There exists some $\xi = \xi(\lambda)$ in (0,1) for which*

$$\lambda \int_{-x_m}^{1} (\hat{a}_\lambda)^2 dx = \int_{-x_m}^{1} (\hat{a}'_\lambda)^2 dx + q_r(\xi;\lambda) \int_{-x_m}^{1} (\hat{a}_\lambda)^2 dx \tag{52}$$

or, in the form of a Rayleigh quotient,

$$\Lambda \int_{-x_m}^{1} (\hat{a}_\lambda)^2 dx = \int_{-x_m}^{1} (\hat{a}'_\lambda)^2 dx, \qquad \textit{with } \Lambda = \lambda - q_r(\xi, \lambda). \tag{53}$$

With the auxiliary conditions (46), it is well known that the minimum value of Λ is

$$\Lambda_s = \left(\frac{\pi/2}{1 + x_m}\right)^2$$

attained when $\hat{a}_\lambda(x)$ is a multiple of the corresponding eigenfunction $\sin(\sqrt{\Lambda_s}(1 - x))$ [19]. It follows that we have

$$\Lambda \geq \Lambda_s \tag{54}$$

since the actual solution $\hat{a}_\lambda(x)$ of (45) and (46) is not $\sin(\sqrt{\Lambda_s}(1 - x))$. In fact, if we let λ_s denote the smallest eigenvalue of the eigenvalue problem (45) and (46). Then we have also

Lemma 4.3. $\Lambda(\lambda_s; \xi) \geq \Lambda_s$.

What we really want to know is the smallest eigenvalue λ_s of the nonlinear eigenvalue problem (45) and (46) which determines the decay rate of transients. Unfortunately, strict inequality generally holds in Lemma 4.3. Even if we have equality instead of inequality, it is still not possible to solve for λ_s because we do not know ξ (which depends on λ_s). Our goal will have to be a more modest one of finding some useful upper and lower bound(s) for the smallest eigenvalue λ_s. The obvious lower bound, $\lambda_s > 0$ (which we know from the previous subsection), is not particularly helpful. More useful bounds have been obtained in [6] by methods similar to those used in [9].

5. Some Steady State Gradients

5.1. *Simplification for $n = 1$*

Whether we take the primary dependent variable as $\bar{a}$ or $\bar{b}$, it is not possible to obtain a useful exact solution of the BVP for the steady state solution in terms of known functions even when the morphogen synthesis rate is piecewise uniform. In this section, we obtain one relatively tractable solution for $n = 1$ to provide some insight to the effect of negative feedback on the steady state morphogen gradients. The qualitative conclusions can be extended to the general case with the help of the theoretical results of Subsection 3.3.

In varying degree of severity, the main obstacle in obtaining any kind of solution when the problem is formulated in terms of $\bar{a}$ is our inability to express $\bar{b}$ in terms of $\bar{a}$. This obstacle does not exist for the case $n = 1$. In that case, we can solve (28), $P(\bar{b}) = 0$, which is now a quadratic equation for $\bar{b}$, to get the following expression for $\bar{b}$ in terms of $\bar{a}$:

$$\bar{b}(x) = b^*(\bar{a}) = \frac{-|\alpha_0 + \varsigma_\gamma \bar{a}| + \sqrt{(\alpha_0 + \varsigma_\gamma \bar{a})^2 + 4\gamma\bar{a}(\alpha_0 + \varsigma\bar{a})}}{2\gamma(\alpha_0 + \varsigma\bar{a})}, \tag{55}$$

with

$$\varsigma_\gamma = \varsigma - \gamma\rho_R = \frac{g_0}{g_r} - \gamma\frac{\bar{V}_{\min}}{\bar{V}_{\max}}. \tag{56}$$

(We have used $|\alpha_0 + \varsigma_\gamma \bar{a}|$ to anticipate the possibility of $\alpha_0 + \varsigma_\gamma \bar{a}$ may be negative for sufficiently large γ.) With (55) and the first relation in (26), equation (25) may then be written as a single ODE for $\bar{a}$ with the help of $\bar{b}(x)$:

$$\bar{a}'' - g_0 b^*(\bar{a}) + v_L(x) = 0 \qquad (-x_m < x < 1)\,. \tag{57}$$

The second order ODE (57) is supplemented by the two boundary conditions (37), keeping in mind also the continuity conditions on $\bar{a}$ and $\bar{a}'$ at $x = 0$.

Note that for sufficiently small γ so that $4\gamma\bar{a}(\alpha_0 + \varsigma\bar{a})/(\alpha_0 + \varsigma_\gamma\bar{a})^2 \ll 1$, we have as a two-term Taylor approximation

$$\bar{b}(x) = b^*(\bar{a}) \sim \frac{\bar{a}}{\alpha_0 + \varsigma_\gamma \bar{a}} \left\{1 - \frac{2\gamma\bar{a}(\alpha_0 + \varsigma\bar{a})}{(\alpha_0 + \varsigma_\gamma\bar{a})^2}\right\}. \tag{58}$$

The first term corresponds to the case of no feedback and the ODE (57) reduces to the corresponding governing ODE in [9] for $\gamma = 0$. The effects of a small γ is characterized by the second term shown the consequence of a small feedback on the receptor synthesis rate.

5.2. *Exact Solution*

For our choice of synthesis rates V_L and V_R, we have $v_L = 0$ in the range $0 < x < 1$ so that

$$\bar{a}'' = g_0 b^*(\bar{a}) \qquad (0 < x < 1)\,. \tag{59}$$

The ODE (59) admits a first integral. The auxiliary condition $\bar{a}(1) = 0$ determine the constant of integration to give

$$[\bar{a}'(x)]^2 - s_1^2 = g_0 I(\bar{a}), \qquad (x > 0) \tag{60}$$

with

$$\frac{dI(\bar{a})}{d\bar{a}} = \bar{b}(\bar{a}), \qquad [I(\bar{a})]_{x=1} = I(0) = 0, \qquad s_1 = \bar{a}'(1). \tag{61}$$

Similarly, we have for the interval $(-x_m, 0)$

$$[\bar{a}'(x)]^2 = g_0[I(\bar{a}) - I(a_m)] \qquad (x < 0), \qquad a_m = \bar{a}(-x_m). \tag{62}$$

In (60) and (62), s_1 and a_m are two unknown constants to be determined by the solution process. The continuity of $\bar{a}'(x)$ and $\bar{a}(x)$ at $x = 0$ requires

$$g_0 I(a_m) + s_1^2 = 0. \tag{63}$$

The first order ODE (60) and (62) can be integrated once more to give

$$\int_{\bar{a}}^{a_0} \frac{da}{\sqrt{I(a) + s_1^2/g_0}} = \sqrt{g_0}x \qquad (0 < x < 1), \qquad a_0 = \bar{a}(0) \tag{64}$$

and

$$\int_{\bar{a}}^{a_m} \frac{da}{\sqrt{I(a) - I(a_m)}} = \sqrt{g_0}(x + x_m) \qquad (-x_m < x < 0). \tag{65}$$

It remains to determine s_1 and a_0 by the continuity condition on $\bar{a}$ at $x = 0$:

$$\int_{a_0}^{a_m} \frac{da}{\sqrt{I(a) - I(a_m)}} = \sqrt{g_0}(x_m), \tag{66}$$

and the end condition $\bar{a}(1) = 0$:

$$\int_{0}^{a_0} \frac{da}{\sqrt{I(a) + s_1^2/g_0}} = \sqrt{g_0}. \tag{67}$$

The relations (63), (67) and (66) are three conditions for determining the three unknown constants a_0, a_m and s_1 to complete the solution process.

5.3. *A Narrow Ligand Source Width*

A significant simplification of the solution can be attained when $X_{\min}$ is small compared to $X_{\max}$. In that case, we may, to a good first approximation, take $X_{\max}$ to be infinite. (Correspondingly, we should use some other reference length X_0, such as $X_{\min}$ or the typical span of the posterior compartment which is about 100 cell deep, instead of $X_{\max}$ in (8) - (11).) For this approximation, the end condition at $x = 1$ is replaced by

$$\lim_{x\to\infty} \bar{a}(x) = 0, \qquad \lim_{x\to\infty} \bar{a}'(x) = 0. \tag{68}$$

The second condition requires $s_1^2 = 0$ so that the condition (63) simplifies to an equation that determines a single unknown a_m:

$$I(a_m) = 0. \tag{69}$$

The condition (66) then becomes an equation for determining a_0 alone.

Even with these simplifications, the values for the two unknown parameters a_m and a_0 must still be obtained numerically by some iterative methods. For each iteration, the values of the two integrals involved must be calculated numerically by an appropriate quadrature formula. Though it is theoretically gratifying to have an exact solution for the problem, the actual solution process is no simpler than to solve the original BVP in the interval $(-x_m, 1)$ directly by some suitable numerical method. For our relatively simple BVP for a single second order ODE of the form $y'' = f(y)$, there exist a number numerical software for its solution in Maple (*desolve*) or MatLab (*bvp4c*). The same results can also been obtained by numerical integration of the original IBVP (12)-(16) in Section 2. Some accurate numerical simulations of our problem will be reported in the next section for the special case of $n = 1$ to provide some insight on the typical effects of our form of negative feedback on the morphogen gradients. They will also serve as background information and motivation for some rigorous results to be obtained there for the general case of $n \geq 1$.

6. Effects of the Negative Feedback

To see the effects of the negative feedback of the form (1), we have implemented numerical simulations using Maple *desolve* (and independently confirmed with Matlab software) for a unit Hill's coefficient, i.e., $n = 1$ in (1), and for a set of typical parameter values in Table I below which was used in [9]. Simulations were implemented also for some positive values of the feedback parameter γ which did not appear in [9].

Table I: System Parameter Values

$X_{\max}$	$X_{\min}$	$r_{\deg}$	$k_{\deg}$	k_{off}	$k_{on}\bar{R}_0$	γ
$0.01cm$	$10^{-3}cm$	$10^{-3}/\sec.$	$2\times10^{-4}/\sec.$	$10^{-6}/\sec.$	$0.01/\sec.$	1.0
						2.0
D	$V_L/\bar{R}_0$	g_r	g_0	f_0	h_0	10.
$10^{-7}cm^2/\sec.$	$10^{-3}/\sec.^*$	1.0	0.2	0.001	10	50.
						∞
$\bar{V}_{\max}$	$\bar{V}_{\min}$	ρ	ρ_R	Δ_R	n	0^*
$10^{-3}/\sec.$	$10^{-4}/\sec.$	1.0	0.1	0.9	1	

*Normal Dpp synthesis rate (of $V_L/\bar{R}_0 = 10^{-3}$/sec.) to be doubled for cases with feedback.

Figure (1) shows a typical comparison of the computed distributions of $b = [LR]/\bar{R}_0$ (for $n = 1$). The lower dashed curve is the distribution for a normal ligand synthesis rate of $V_L/\bar{R}_0 = 10^{-3}/\sec.$ without feedback ($\gamma = 0$). The upper dashed curve is a similar distribution with the ligand synthesis rate doubled (again without feedback). Naturally, the concentration of the normalized bound morphogen is much higher for the latter case since *Dpp* is produced at a higher rate. When negative feedback of the form (1) is introduced with $\gamma = 1$ for the higher synthesis rate case, the distribution of bound morphogens is changed to the solid curve. While $[LR]/\bar{R}_0$ is reduced near the ligand source to a level close to that before doubling the ligand synthesis rate, the shape of the gradient curve is less convex, in fact slightly concave in the important range of $(0, 1)$ where the original gradient is appropriately convex.

Figure (2) shows computed distributions of $[LR]/\bar{R}_0$ across the posterior compartment for selected values of γ (and $n = 1$). Of the two solid curves, the one that

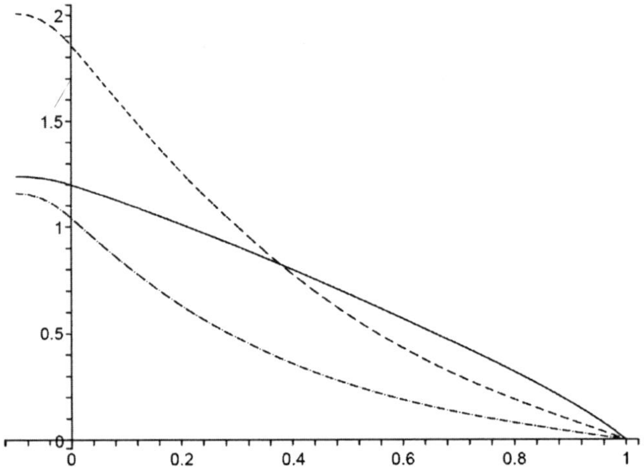

Figure 1. $[LR]/\bar{R}_0$ vs. x for different combination of $\{\gamma, \bar{v}_L\}$ with the remaining parameters values given in Table 1. The dash-dot curve is for $\{0, 1\}$, the dash curve is for $\{0, 2\}$, and solid curve is for {1,2}.

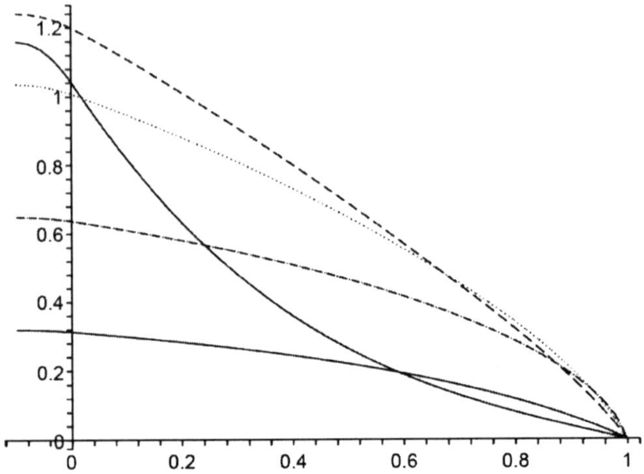

Figure 2. $[LR]/\bar{R}_0$ for $\gamma = 0$ with $\bar{v}_L = 1.$ and for $\gamma = 1,\ 2,\ 10,\ 50, \infty$ (= upper solution) with $\bar{v}_L = 2$ in all five cases. All other parameter values are as given in Table 1.

is markedly convex in the interval $(0, 1)$ is one for the normal ligand synthesis rate without feedback. The corresponding signaling ligand concentrations after doubling ligand synthesis rate and adding feedback with $\gamma = 1, 2, 10$ and 50 are shown as the four (dash and dot) curves in the figure. The magnitude of the signaling ligand concentration near the source region is lower for higher γ. However, the situation is more complex near the sink edge. While the magnitude of $[LR]/\bar{R}_0$ increases slightly as γ increases from 1 to 10, the increase over the previous level occurs in a smaller and smaller interval adjacent to the edge $x = 1$. For still larger γ (such as

$\gamma = 50$), $[LR]/\bar{R}_0$ actually decreases with γ for the entire posterior compartment, approaching in the limit as γ tends to ∞ the solution for $\bar{b} = B/A$ (obtained by taking $\gamma = \infty$ in (28)) given by the lower solid (concave) line curve in the figure.

For the $n = 1$ case, the shape of the signaling gradient for any positive γ is generally less convex and more concave than the corresponding gradient without feedback ($\gamma = 0$). Though the reduction in convexity is not monotone in γ near the sink edge, the general increase in concavity over the $\gamma = 0$ case indicates that our negative feedback on receptor synthesis rate leads to a leveling effect on the signaling morphogen gradient. The change in gradient convexity results in undesirable deviations from the target gradient for the original biological pattern and does not promote robustness in development.

The qualitative effects of a low level negative feedback of the form (1) for $n = 1$ can also be seen from (58) for a sufficiently small γ (not shown in the two figures). Had the free *Dpp* gradient $\bar{a}(x)$ remained unchanged, the (sufficiently small) negative feedback would reduce the magnitude of the bound morphogen gradient $\bar{b}(x)$ for all x in $(-x_m, 1)$. While this generally contributes to the robustness of the development of the wing imaginal disc, this positive effect is mediated by the differential reduction induced by the spatial variation of the free *Dpp* concentration. With $\bar{a}(x)$ monotone decreasing from $-x_m$ to 1, the percentage reduction of $\bar{b}(x)$ is most severe near the morphogen source and much less substantial adjacent to the wing disc edge. While $\bar{b}(x)$ would remain convex for sufficiently small γ, this differential magnitude adjustment has the net effect of leveling the gradient in addition to magnitude reduction.

However, the presence of the negative feedback does in fact alter the steady state free *Dpp* concentration gradient and thereby complicates its actual effect on the final steady state signaling (bound) *Dpp* gradient responsible for differential cell expression. To the extent that the negative feedback does not alter the monotone decreasing property of $\bar{a}(x)$, the leveling of the bound *Dpp* gradient by a negative feedback of the form (1) continues to operate, differing only in the degree of severity. Hence, for sufficiently small values of γ, the particular form of negative feedback (1) tends to reduce cell expression differentially, higher reduction (for the lower receptor synthesis rate) near the source and lower (for the higher receptor synthesis rate) near the edge. Similar to the moderate and high γ values, this leveling effect of a low level negative feedback also changes the biologically desirable convexity of bound morphogen gradient and thereby works against robustness of wing disc development.

These observations (supported by results of numerical solutions of the relevant BVP and by the approximate formula (58)) provide an explanation for the negative simulation results (that feedback of the form (1) does not improve robustness of the biological development of the Drosophila wing imaginal disc). Clearly, a reduction in receptor synthesis rate leads to more *Dpp* available for downstream binding with signaling receptors. However, the differential reduction of *Dpp* synthesis rate induced by a spatially nonuniform bound morphogen concentration changes $\bar{b}(x)$ differentially in space, higher reduction near the source end and lower reduction near the sink edge,resulting in a less convex and (for a sufficiently large γ) possibly concave $\bar{b}(x)$. We will now validate rigorously the aforementioned explanation for the negative simulation results first reported in [12] for the $n = 1$ case for the entire range of γ values and extend it to the more general case of $n > 1$.

By Proposition 3.5, the steady state free *Dpp* concentration gradient for a fixed γ is always monotone decreasing as x increases. By Lemma 3.2, the change in steady state bound morphogen gradient $\bar{b}(x)$ for a fixed γ is also monotone decreasing for increasing x as a consequence. With a spatially nonuniform $\bar{b}(x)$, the leveling effect of our form of negative feedback therefore persists for the general case. However, there are some qualitative differences in this effect for γ not sufficiently small so that (58) is not an accurate characterization of the effects of (1). The following results provide a rigorous and more complete characterization of the effect of our negative feedback for a general Hill's coefficient.

Proposition 6.1. *$\bar{a}(x;\gamma)$ is an increasing function of γ for $-x_m < x < 1$.*

Proof. Let $w(x) = \partial\bar{a}/\partial\gamma$. Upon differentiating the ODE (36) and BC (37) for $\bar{a}$ partially with respect to γ, we obtain with the help of (28)-(30)

$$-w'' + g_0\sigma_a w - g_0\sigma_\gamma = 0 \qquad (-x_m < x < 1) \tag{70}$$

$$w'(-x_m) = 0, \qquad w(1) = 0 \tag{71}$$

with

$$\sigma_a \equiv \frac{\partial b^*(\bar{a},\gamma)}{\partial\bar{a}} = \frac{A_0 b^*}{\Delta\bar{a}} \geq 0, \tag{72}$$

$$\sigma_\gamma \equiv -\frac{\partial b^*(\bar{a},\gamma)}{\partial\gamma} = \frac{1}{\Delta}\frac{(b^*)^n C}{1+\gamma(b^*)^n} \geq 0, \tag{73}$$

$$\Delta = A[1+\gamma(b^*)^n] + \frac{n\gamma(b^*)^{n-1}C}{1+\gamma(b^*)^n} \tag{74}$$

where $A(\bar{a})$ and $C(\bar{a})$ are as defined in (29) and (30). Evidently, $w_\ell(x) = 0$ is a lower solution of the BVP for w. For an upper solution, let $\sigma_{\max}$ be the maximum of $\sigma_\gamma(x)$ and $w_u(x)$ be the solution of the BVP

$$-w_u'' + g_0\sigma_a w_u - g_0\sigma_{\max} = 0 \qquad (-x_m < x < 1) \tag{75}$$

$$w_u'(-x_m) = 0, \qquad w_u(1) = 0 \tag{76}$$

Then $w_u(x)$ is an upper solution of the BVP for $w(x)$. It follows that there exist a unique $w(x)$ that is non-negative. Furthermore $w(x) \equiv 0$ does not satisfy the ODE (70). Hence, the proposition is proved. □

For $\bar{b}(x,\gamma)$, we obtain from (28)-(30)

$$\frac{\partial \bar{b}(x,\gamma)}{\partial \gamma} = \frac{g_r(f_0+g_0)}{\Delta(1+\gamma\bar{b}^n)}\frac{\bar{b}}{\bar{a}}\left\{(1+\gamma\bar{b}^n)^2\frac{d\bar{a}}{d\gamma} - (1-\rho_R)\frac{\bar{a}^2\bar{b}^{n-1}}{\alpha_0}\right\} \tag{77}$$

where Δ is as defined in (74) and where the ratio $\bar{b}/\bar{a}$ has a finite limit as $x \to 1$:

$$\lim_{x\to 1}\frac{\bar{b}(x,\gamma)}{\bar{a}(x,\gamma)} = \frac{B_1+C_1}{A_0} = \frac{1}{\alpha_0}.$$

Unlike the special case of small γ and $n = 1$, the effects of the feedback parameter γ in (1) on the signaling gradient now depends on the sign of the right hand side of (77).

In the source region and nearby, numerical results for the case of $\gamma = 1, 2, 10$ and 50 all lead to a negative value for $\partial\bar{b}(x,\gamma)/\partial\gamma$. Adjacent to the edge sink at $x = 1$, the addition of a negative feedback on receptor synthesis rate of the form (1) gives a positive value for right hand side of (77) for $\gamma = 1, 2$ and 10 with the interval for the positive value shrinking with increasing γ. For $\gamma > 50$, the region of positive right hand side of (77) is imperceptible, consistent with the numerical simulation results showing that beyond a certain critical value of γ, the signaling morphogen concentration decreases monotonically and pointwise toward the limiting distribution corresponding to the case of $\bar{b} = B(\bar{a})/A(\bar{a}) = B_1\bar{a}/[A_1\bar{a} + A_0]$. The gradient of this limiting solution is still concave for the $n = 1$ case though somewhat less so compared to those for (finite but) large values of γ. As such the addition of a negative feedback of the form (1) drives the signaling gradient away from its target shape for biological development.

This is not to say that negative feedback mechanisms are generally not effective for robustness. But if robustness is to be achieved, the feedback process should not promote a leveling effect on the convex signaling gradient. Rather, our observations suggest that an effective negative feedback mechanism for robustness through regulating the receptor synthesis rate should be spatially uniform. For example, the desired spatial uniform feedback effect may be attained with a negative feedback on the receptor synthesis rate of the form

$$V_R = V_{\min} + \frac{V_{\max}-V_{\min}}{1+\Gamma([LR])} = V_{\max}\left\{\rho_R + \frac{\Delta_R}{1+\gamma(\bar{b}(x))}\right\} \tag{78}$$

where $\gamma(\bar{b}(x))$ is a functional of the signaling gradient $\bar{b}(x)$. Possible choices of $\gamma(\bar{b}(x))$ include $\gamma_0\bar{b}(0)$ and some average value of the signaling morphogen concentration,

$$\gamma(\bar{b}(x)) = \gamma_0\int_0^1 \bar{b}(x)dx. \tag{79}$$

The effects of such spatially uniform feedback mechanisms are being investigated.

7. Conclusion

The present research was motivated by the results of [12] on negative feedback as an instrument for achieving robustness in biological developments. The basic models for morphogen gradient formation that accounts only for diffusion, reversible binding with renewable signaling receptors, and degradation (with or without transcytosis of bound morphogens) were found to be sensitive to changes in system parameters. Doubling the morphogen synthesis rate changes the corresponding signaling bound *Dpp* gradient substantially, exceeding the acceptable range of a robustness measure introduced in [12] for all sets of parameter values simulated. Conventional wisdom would have the sensitivity to system characteristics lowered by some form of negative feedback which is expected to ameliorate the response to the system parameter changes. It was therefore rather surprising that a negative feedback mechanism of the form (1) known to be effective in other applications was found not to render the development of the Drosophila wing imaginal disc more robust [12]. In fact, numerical simulations for 10^6 sets of system parameter values showed generally a higher sensitivity to a doubling of *Dpp* synthesis rate than the basic models. The unexpected finding prompted the present examination of the cause of the ineffectiveness of the negative feedback mechanism (1).

From the approximate analysis and numerical simulations for the special case of $n = 1$, we found that in addition to reducing the receptor synthesis rate and thereby lowering the signaling bound morphogen concentration near the source end, the particular feedback mechanism (1) has a leveling effect on the (normalized) signal bound morphogen gradient $\bar{b}(x) = [LR]/\bar{R}_0$ which works against robustness. The manner in which the particular feedback mechanism changes the slope and convexity of the bound morphogen distribution drives the signaling gradient away from the appropriate gradient shape for the targeted tissue pattern. It would not serve the intended purpose of ensuring robustness to system characteristics changes.

The observations for the special case above were then made rigorous and extended to the general case of $n \geq 1$. For this extension, the results in Lemma 3.2 and monotonicity result of Proposition 3.5 played a key role. This reason alone requires the development of the basic existence theory for the basic model with the negative feedback (1) as presented in Section 2.

The cause of the negative results of [12] suggests more appropriate negative feedback mechanisms for mediating the sensitivity to system parameter changes and thereby ensuring robustness. Such mechanisms should not cause a substantial change in the signaling gradient shape, either by leveling the monotone decreasing gradient or steeping it and making it more convex. Some possible feedback mechanisms for this purpose are currently being investigated.

Acknowledgments

The research of F.Y.M. Wan has supported in part by NIH grants P20GM66051, R01GM067247and R01GM075309. The two NIH R01 grants were awarded through the Joint NSF/NIGMS Initiative to Support Research in the Area of Mathematical Biology. The research project was motivated by and has benefitted from the earlier work by the second author jointly with his UCI colleagues A.D. Lander and Qing Nie on robustness by numerical simulations.

References

1. Amann, H., "On the existence of positive solutions of nonlinear boundary value problems," Indiana Univ. Math. J., Vol. 21, 1971, 125-146.
2. Entchev, E.V., Schwabedissen, A. and Gonzalez-Gaitan, M., "Gradient formation of the TGSF-beta homolog Dpp," Cell Vol. 103, 2000, 981-991.
3. Gurdon, J.B., and Bourillot, P.Y., "Morphogen gradient interpretation," Nature, Vol. 413, 2001, 797-803.
4. Hill, A.V., "The combinations of haemoglobin with oxygen and with carbon-monoxide," I. J. Physiol., vol. 40 (iv-vii), 1910.
5. Kerszberg, M. and Wolpert, L., Mechanisms for positional signalling by morphogen transport: a theoretical study," J. Theor. Biol., Vol. 191, 1998, 103-114.
6. Khong, M., "Feedback and Morphogen Gradients," Ph.D. dissertation research, University of California, Irvine, in progress.
7. Lander, A.D., Nie, Q., Vargas, B. and Wan, F.Y.M., "Aggregation of Distributed Sources in Morphogen Gradient Formation," J. Comp. Appl. Math., Vol. 190, 2006, 232-251.
8. Lander, A.D., Nie, Q. and Wan, F.Y.M., "Do Morphogen Gradients Arise by Diffusion?" Developmental Cell, Vol. 2, 2002, 785-796.
9. Lander, A.D., Nie, Q. and Wan, F.Y.M., "Spatially distributed morphogen production and morphogen gradient formation," Math. Biosci. Eng. (MBE), Vol. 2 , 2005, 239 – 262.
10. Lander, A.D., Nie, Q. and Wan, F.Y.M., "Internalization and end flux in morphogen gradient formation," J. Comp. Appl. Math 2006, 232-251.
11. Lander, A.D., Nie, Q., Wan, F.Y.M., and Xu, J., "Diffusion and Morphogen Gradient Formation - Part I: Extracellular Formulation," submitted to J. Math. Bio.
12. Lander, A.D., Wan, F.Y.M., and Nie, Q., "Multiple Paths to Morphogen Gradient Robustness," 2005, submitted for publication.
13. Lou, Y., Nie, Q. and Wan, F.Y.M., "Nonlinear eigenvalue problems in the stability analysis of morphogen gradients," Studies in Appl. Math., Vol. 113, 2004, 183-215.
14. Sattinger, D.H., "Monotone Methods in Nonlinear Elliptic and Parabolic Boundary Value Problems," Indiana University Math. J., Vol. 21, 1972, 981-1000.
15. Smoller, J., *Shock Waves and Reaction-Diffusion Equations*, Springer Verlag Inc., New York, 1982.
16. Teleman, A.A., and Cohen, S.M., "Dpp gradient formation in the Drosophila wing imaginal disc," Cell, Vol. 103, 2000, 971-980.
17. Vargas, B., *Leaky boundaries and morphogen gradient formation*, Ph.D. Dissertation, University of California, Irvine, December, 2006.
18. Lander, A.D., Q. Nie, Q., Vargas, B. and Wan, F.Y.M., "Apical-Basal Cell Depth and Morphogen Robustness," to appear.
19. Wan, F.Y.M., *Introduction to the Calculus of Variations and Its Applications*, Chapman & Hall, New York, 1995.

CSAW: STOCHASTIC APPROACH TO PROTEIN FOLDING

KERSON HUANG
Physics Department, Massachusetts Institute of Technology,
Cambridge, MA 02139, USA
E-mail: kerson@mit.edu

CSAW (conditioned self-avoiding walk) is a model of protein folding that combines the features of SAW (self-avoiding walk) and the Monte-Carlo method. It simulates the Brownian motion of a chain-molecule in the presence of interactions. The model is designed so that one can add desired features step by step. It can be used as guide to strucuture prediction, as well as theoretical laboratory.

1. Introduction

C.C. Lin, whose 90th birthday we are honoring, has emphasized that protein folding is a stochastic dissipative process [1]. I want to describe CSAW (conditioned self-avoiding walk) [2], a model that reflects that philosophy, and report on some results.

The plan of this talk is as follows. I will begin with a brief review of background facts on protein folding. I will then describe SAW (Self-avoiding walk), which simulates the unfolded state of a protein. The folding process is then simulated by adding conditions representing interactions, hence CSAW. I will illustrate the method by a series of examples, starting from the very simple to a realistic simulation. I will also comment on the use of this model as a theoretical laboratory.

2. Protein folding

A protein is a chain molecule that performs biological function in living cell. The nucleus of a cell contains chromosomes, which consists of tightly wound strands of DNA, along which are segments called genes. Each gene contains the code to make one protein.

The protein chain consists of a sequence of units called "residues", which are amino acids chosen from a pool of 20. This sequence is called the *primary structure*. The amino acids differ only in the side chain, as illustrated in Fig.1. The center of each amino acid is a carbon atom labeled C_α. In the lower panel of Fig.1, we depict the bare skeleton of the protein as a sequence of C_α's connected by chemical bonds, omitting all other atoms. The connecting bonds form a rigid "crank" that lies in a plane. The relative orientation of two successive cranks are specified by two angles. For the purpose of protein folding, the only degrees of freedom of the protein chain are these "torsional angles". A protein of N residues therefore has $2(N-1)$ degrees

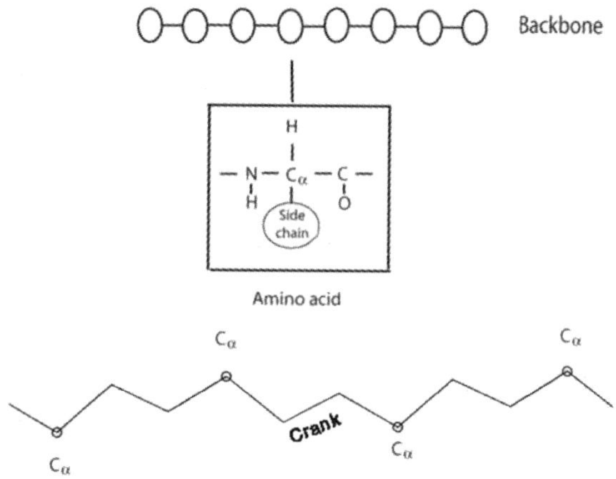

Figure 1. Skeleton of a protein can be depicted as a series of carbon atoms connected by cranks. The only degrees of freedom are the relative angles between planes of the cranks.

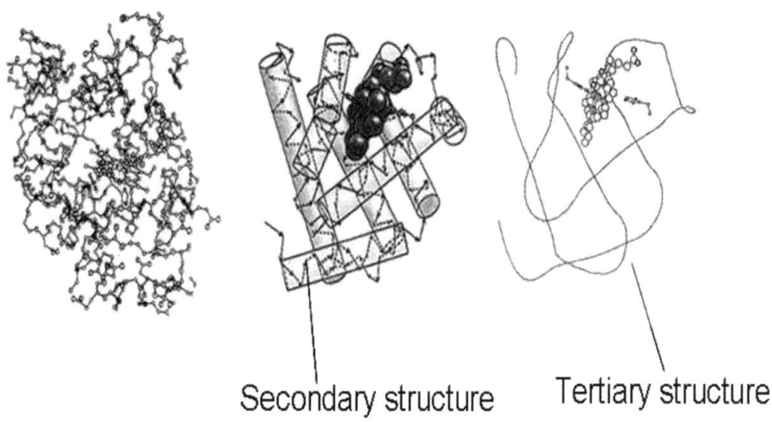

Figure 2. Native state of myoglobin, showing different degrees of detail.

of freedom.

In a non-aqueous solution:the protein is in an unfolded state that can be represented by a random coil. In water it folds into the "native state" with a definite geometrical shape. Fig.2 shows the native state of myoglobin, with different degrees of detail. Local structures, such as helices, are called *secondary structures.* When they are omitted, one sees the 3D outline called the *tertiary structure.* Native state of myoglobin, showing different degrees of detail. Secondary structures are of two manly types, the alpha helix and the beta sheet, as shown in Fig.3. The former is stabilized by hydrogens bonds, which connects atoms of residues 1 with 4, 2 with

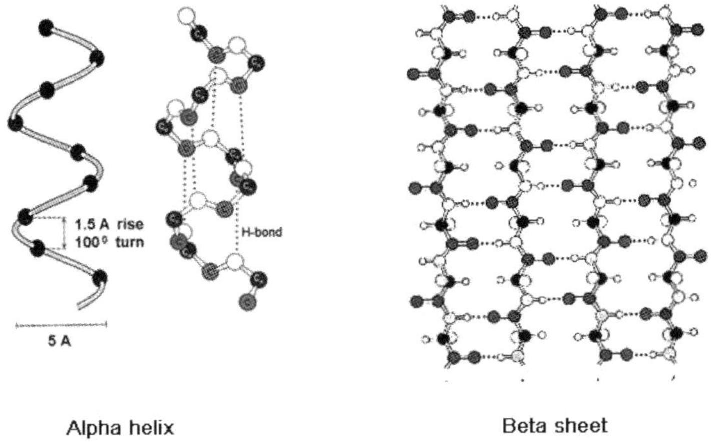

Figure 3. Secondary structures: alpha helix and beta sheet. Hydrogen bonds are shown as dotted lines.

5, *etc.* The beta sheet is a mat sewn together by hydrogen bonds.

The living cell is mostly water, whose molecules form a network of hydrogen bonds, which switch between different molecules on a time scale of 10^{12}s. Foreign molecule are "soluble", or "hydrophilic", if they can hydrogen-bond with water. Otherwise they are "insoluble", or "hydrophobic". On the protein chain, there are both hydrophilic and hydrophobic side chains. When immerse in water, the protein folds in order to shield the hydrophobic side chains from water. This is called the "hydrophobic effect". In this process, a frustration arises, because the skeleton is hydrophilic, as illustrated in Fig.4. The resolution is the formation of secondary structure, which uses up the skeletal hydrogen bonds internally.

As depicted schematically in Fig.4, a typical folding process consists of a very rapid collapse into an intermediate state called the "molten globule", which lasts for a relatively long time before making a final transition to the native state. The collapse time is generally less than 200 μs, while the molten globule, which is almost as compact as the native state, can last as long as 10 minutes.

The biological function of the protein depends on its native shape. The cup-like shape of the protein shown in Fig.5, for example, provides a binding site for a specific molecule. Models of protein folding

Model of protein folding generally fall into two types: bioinformatics and computations. In the former, one tries to cull empirical rules on the relation between primary sequence and native state from the database of known proteins. In the latter, one tries to compute the coordinates of the atoms in the native state by various schemes. The former is devoid of dynamical content, while the second deals with atomic interactions to varying degrees, depending on the assumptions. Ideally, one would like to not only predict the native state from the primary sequence, but

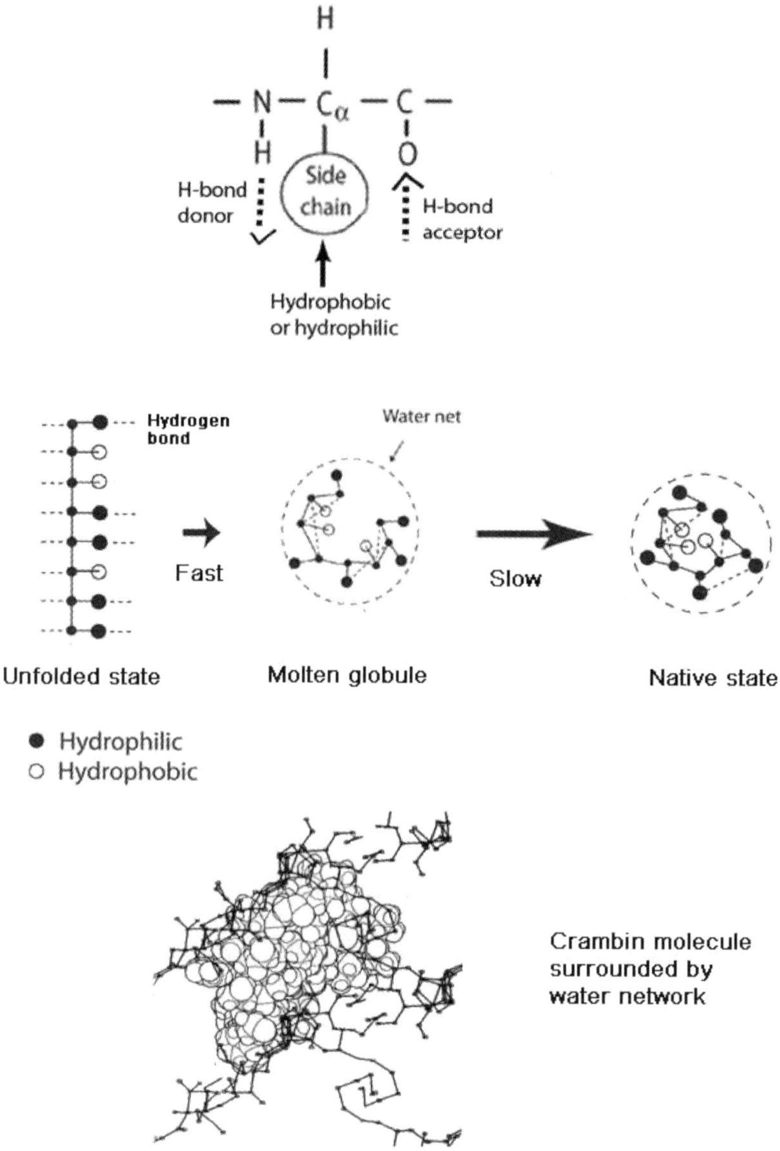

Figure 4. Upper panel: Side chain may be hydrophilic or hydrophobic, but atoms on main chain are hydrophilic. Middle: Schematic picture of folding process. Lower: Protein is "squeezed" into shape by water network.

also understand the physical principles behind the folding process. So far, one can achieve a 75% success rate in some predictions, but there is woefully little physical understanding.

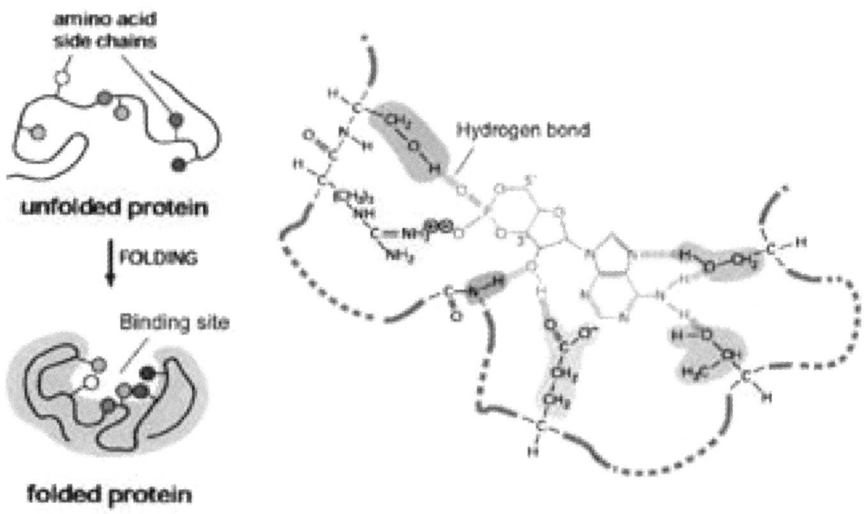

Figure 5. Native shape tailored to the binding of a specific molecule.

A popular method of computation is molecular dynamics (MD). This is a "brute force" approach that solves Newton's equation on a computer, for the motions of all the atoms in the protein, plus thousands of water molecules, using known potenitals between all pairs of atoms present. The best effort so far can following the folding of a medium-sized protein to 1 μs, whereas the actual folding process requires minutes. In the time span of the calculation, the protein has not even completed the initial fast collapse. Furthermore, this methods does not provide insight into the mechanism of folding.

Describing the water molecule by molecule seems to be a stupendous waste of computing resources. Instead, we propose a method based on Brownian motion. The unfolded chain is simulated by self-avoiding walk (SAW), a random walk in which the residues are prohibited from overlapping each other. The folding process is then modeled by adding extra conditions, in order to represent the hydrophobic effect and interactions such as hydrogen bonding. These conditions are implemented by selecting an appropriate subensemble from the ensemble of SAW, through a Monte-Carlo procedure. This is the conditioned self-avoiding walk (CSAW).

3. Brownian motion

Let us review the Brownian motion of a single particle suspended in a medium. Its position $x(t)$ is a stochastic variable describe by the Langevin equation [3]

$$m\ddot{x} = F(t) - \gamma\dot{x}. \tag{1}$$

The force on the particle by the medium is split into two parts: the damping force $-\gamma\dot{x}$ and a random component $F(t)$, which belongs to a statistical ensemble with the properties

$$\langle F(t)\rangle = 0$$
$$\langle F(t)F(t')\rangle = c_0\delta(t-t') \tag{2}$$

where the brackets $\langle\rangle$ denote ensemble average. The equation can be readily solved. What we want to emphasize is that it can also be simulated by random walk [4]. Both procedures give rise to diffusion, with diffusion coefficient $D = c_0/\left(2\gamma^2\right)$.

4. Monte-Carlo

In the presence of a regular (non-random) external force $G(x)$, the Langevin equation may not be soluble analytically , but we can solve it on a computer via *conditioned random walk.* We first generate a trial step at random, and accept it with a probability according to the Monte-Carlo method. Let E be the potential energy of the external force G. Let ΔE be the energy change in the proposed update. The algorithm is as follows:

- If $\Delta E \leq 0$, accept the proposed update;
- If $\Delta E > 0$, accept the proposed update with probability $\exp\left(-\Delta E/k_B T\right)$.

The last condition simulates thermal fluctuations. After a sufficiently large number of updates, the sequence of state generated will yield a Maxwell-Boltzmann distribution with potential energy E.

The conditioned random walk may be described in terms of the Langevin equation as follows:

$$m\ddot{x} = \underbrace{[F(t) - \gamma\dot{x}]}_{\text{Treat via random walk}} + \underbrace{G(x).}_{\text{Treat via Monte-Carlo}} \tag{3}$$

5. SAW

SAW can be simulate via the pivot algorithm [4] [5], as follows. Choose an initial chain configuration in 3D continuous space, and hold one end of the chain fixed.

- @ Choose an arbitrary point on the chain as pivot.
- Rotate the end portion of the chain rigidly about the pivot (by changing the torsional angles at the pivot point).
- If this does not result in any overlaps accept the configuration as an update, otherwise go to @..

By repeating this procedure, one eventually generates a uniform ergodic ensemble of SAW's, which simulates a generalized Langevin equation of the form

$$m_k\ddot{\mathbf{x}}_k = \mathbf{F}_k(t) - \gamma_k\dot{\mathbf{x}} + \mathbf{U}_k(\mathbf{x}_1\cdots\mathbf{x}_N), \quad (k = 1\cdots N) \tag{4}$$

where the subscript $k = 1 \cdots N$ labels the residue along the chain. The term $\mathbf{U}_k$ includes the forces that maintain the bonds between successive residues, and prohibit the residues from overlapping one another.

6. CSAW

We can include the hydrophobic effect, and other interactions such as hydrogen-bonding, by adding a force term:

$$m_k \ddot{\mathbf{x}}_k = \underbrace{(\mathbf{F}_k - \gamma_k \dot{\mathbf{x}} + \mathbf{U}_k)}_{\text{Treat via SAW}} + \underbrace{\mathbf{G}_k.}_{\text{Treat via Monte-Carlo}} \qquad (k = 1 \cdots N) \qquad (5)$$

Treating the forces $\mathbf{G}_k$ by the Monte-Carlo method results in CSAW.

7. Hydrophobic model

We take the residues as impenetrable hard spheres, with diameter equal to the distance between successive residues. The hydrophobic effect is taken into account using a simple "PH model" [6] in which the residues are either hydrophilic (P) or hydrophobic (H). Hydrogen-bonding is not taken into account at this stage.

The centers of the hard spheres (corresponding to C_α atoms,) are connected by the cranks described before. The energy E of a configuration is given by

$$\begin{aligned} E &= -gK \\ K &= \text{Contact number of hydrophobic residues} \end{aligned} \qquad (6)$$

A residue's contact number is the number of residues in contact with it, *not counting those lying next to it along the chain.* Two residues are in contact when their centers are separated by less than a set fraction (say, .1.2) of the residue diameter. The unfolded chain in a non-aqueous solution corresponds to $g = 0$. When two H's are in contact, the total contact number increases by 2. This induces an attractive force between H residues.

We first illustrate the model in a simple setting. Consider a chain of 7 circular monomers in 2D continuous space, with only one H. The monomers are connected by straight bonds. We know that the configuration for maximal shielding of H from the medium is a hexagonal close pack with H at the center. Results from CSAW are shown in Fig.6, which indeed reproduce the hexagonal close pack.

Simulations of a 3D chain of hard spheres connected by cranks are shown in Fig.7, with one H and two H's. Close-packing of free spheres in 3D will lead to hexagonal or cubic close pack, in which each sphere touches 12 nearest neighbors [7]. This cannot be realized here, because of the intervention of the connecting cranks. However, a hexagonal geometry can be discerned. The case with two H's shows an effective attraction.

Fig.8 shows the results for a chain of 30 cranks, with one third of the residues being hydrophobic. The chain settles into a final structure resembling a three-leaf

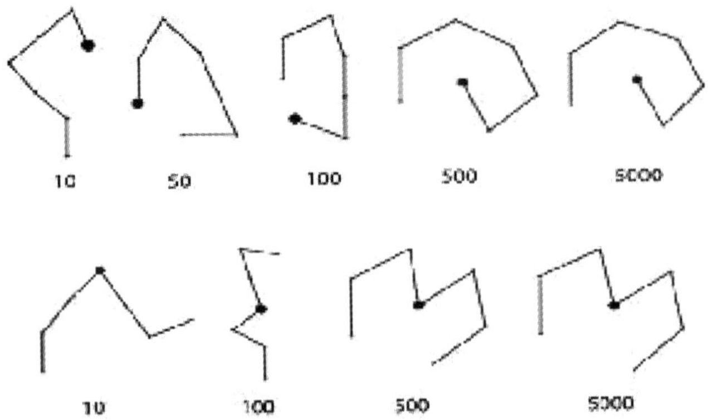

Figure 6. 2D illustration of CSAW. The black dot represents the hydrophobic residue, Numbers correspond to Monte-Carlo steps.

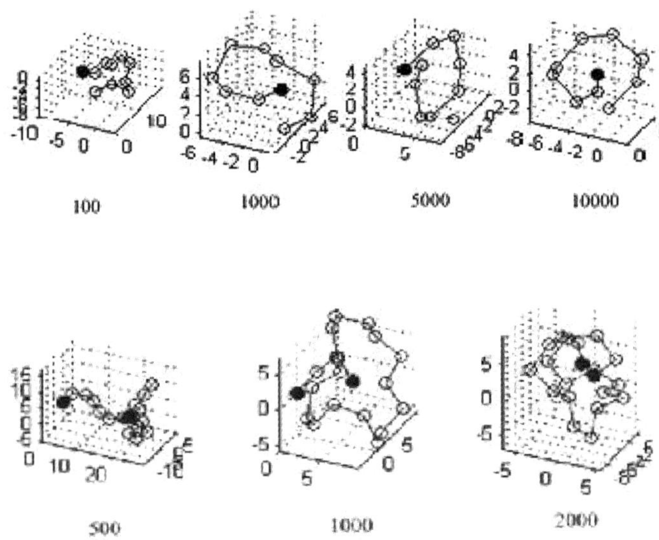

Figure 7. Folding of hard-sphere chains in 3D, with one and two hydrophobic residues (black dots).

clover. From these results, we conclude that hydrophobic forces alone leads to a well-defined tertiary structure in the final state.

8. Hydrogen bonding

We now add hydrogen bonding. To do this, we attach the O and N atoms to the crank, so that there are $C = O$ and $N - H$ bonds in the plane of the crank, as shown

Figure 8. Folding a 3D chain of 30 residues, one third of which are hydrophobic.

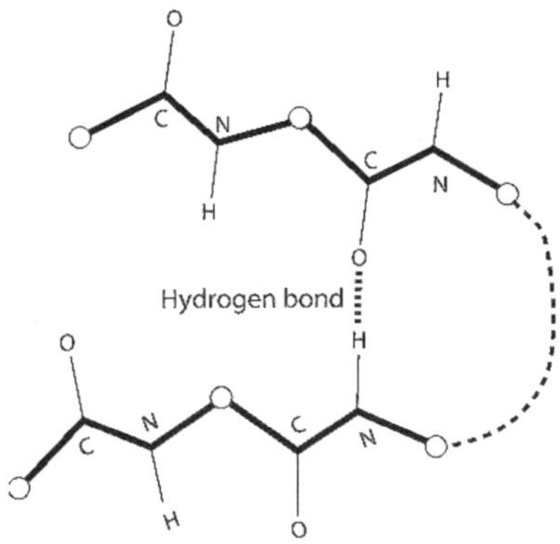

Figure 9. A hydrogen bond is formed when the distance between O and H lie within given limits, and the bonds they belong to are antiparallel within given tolerance.

in Fig.9. A hydrogen bond is formed between O and H from different residues, when

- the distance between O and H is 2 A, within given tolerance;
- The bonds $C = O$ and $N - H$ are antiparallel, within given tolerance.

The residues are still either hydrophobic (H) of hydrophilic (P); but they are no longer treated as hard spheres. The self-avoidance condition now takes into account collisions among the H and O atoms on the newly added bonds. The energy of a configuration is now

$$\begin{aligned} E &= -g_1 K_1 - g_2 K_2 \\ K_1 &= \text{Contact number of hydrophobic residues} \\ K_2 &= \text{ Number of hydrogen bonds} \end{aligned} \tag{7}$$

As a test, we turn off the hydrophobic interaction by setting $g_1 = 0$. With $n = 30$, $g_2 = 1$, an α-helix emerges after 1000 Monte-Carlo steps, as depicted in Fig.10.

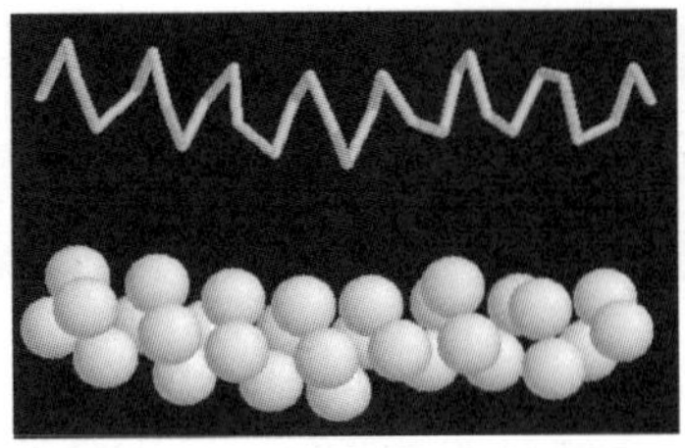

Figure 10. Alpha helix generated by the hydrogen bonding interaction. In this test run the hydrophobic interaction is turned off.

9. Molten globule

In the absence of hydrogen bonding, the folding chain did not go through an intermediate state, but collapses to the final state. By including hydrogen bonding, we can reproduce the molten globule. We consider a chain of 30 residues, 10 of which are hydrophobic. The evolution of the radius of gyration is shown in Fig.11. After a rapid collapse, there is a long-lived metastable state that we can identify with the molten globule. The transition to the native state mimics experimental data from the folding of apomyoglobin [8], as displayed in the lowest panel of Fig.11.

Fig.12 shows chain configurations. The native state consists of two short helices connected by a loop.

10. Folding pathways of chignolin

We now consider a real protein, chignolin, which has 10 residues. All atoms are modeled realistically, except that the side chains are approximated by hard spheres. The native state emerges after 70000 steps, as shown in Fig.13. The computation, done by Dr. Jinzhi Lei of Tsinghua University, took less than 5 minutes on a work station. In contrast, an MD simulation on the same work station did not reach the native state in one month's computation. Details of the folding process for 100 runs are summarized in Figs.14-16.

11. Correlation functions

The protein energy spectrum is important for understanding its interaction with water. As shown in Fig. 17, there is a resonance between the normal modes of the water network and a protein molecule.

We can study protein modes by calculating various correlation functions via

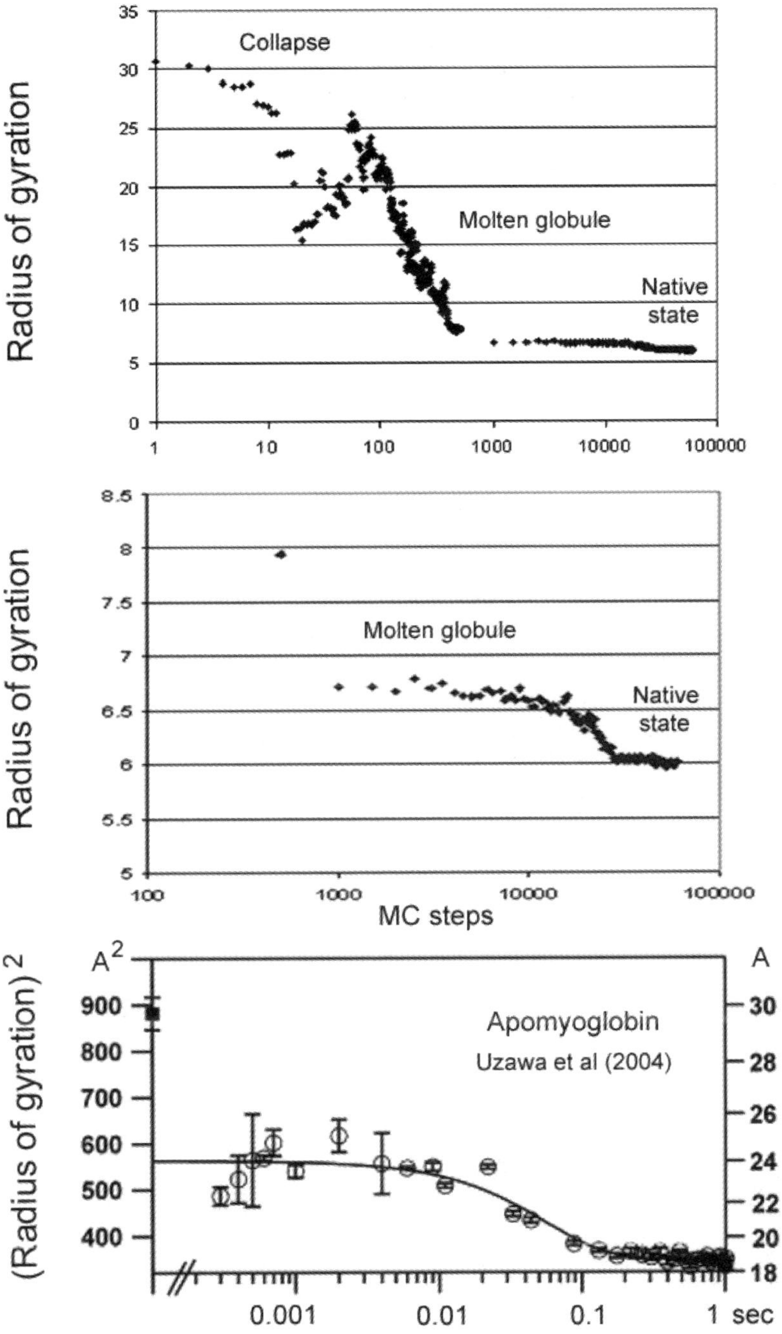

Figure 11. Upper panel: Radius ofgyration. Middle:Magnified view of the final transition to the native state. Lower: Experimental data from the folding of apomyoglobin, by group at Osaka Univerity [8].

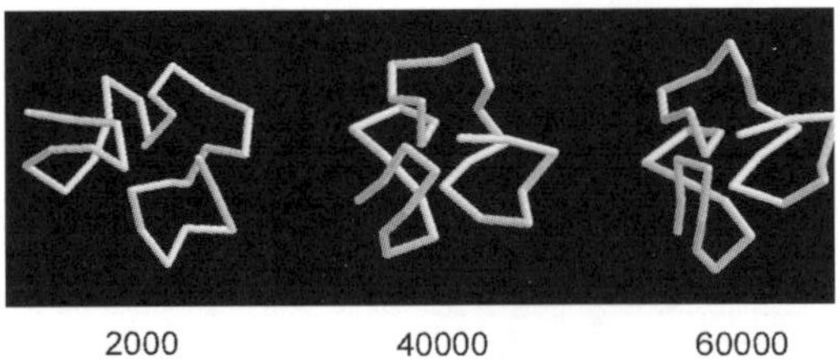

Figure 12. Chain of 30 residues, 1/3 of which are hydrophobic. The first two images corresponds to the molten globule state, and the last corresponds to the native state.

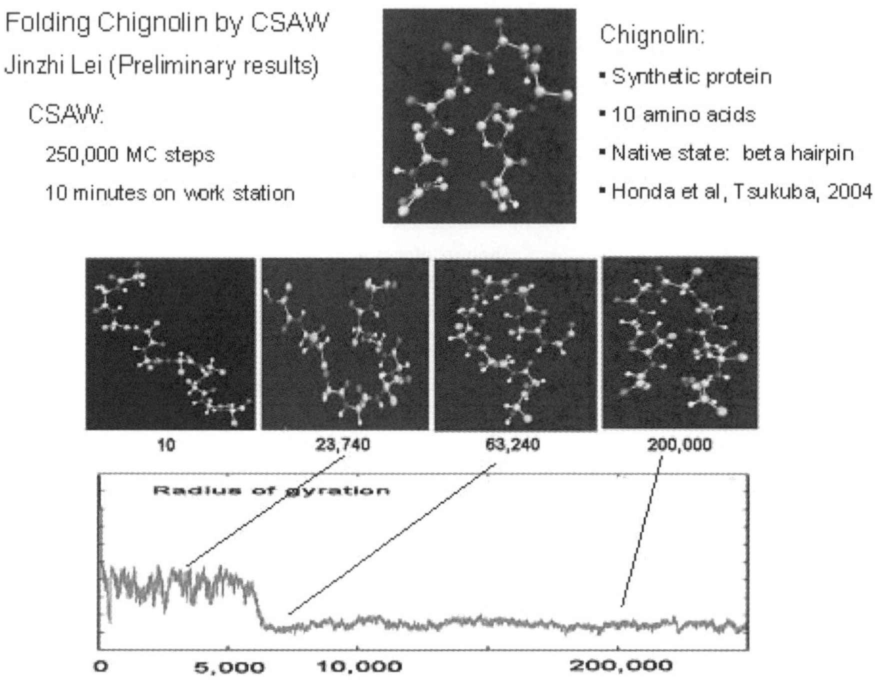

Figure 13. Folding chignolin by CSAW.

CSAW. Some relevant functions are

$$E(\mathbf{k}) = \int d\mathbf{x} e^{i\mathbf{k}\cdot\mathbf{x}} \langle \mathbf{J}(\mathbf{x}) \cdot \mathbf{J}(0) \rangle \qquad \text{(Energy spectrum)}$$

$$s(\mathbf{k}) = \int d\mathbf{x} e^{i\mathbf{k}\cdot\mathbf{x}} \langle \rho(\mathbf{x}) \cdot \rho(0) \rangle \qquad \text{(Static structure factor)}$$

$$S(\mathbf{k}, \omega) = \int d\mathbf{x} e^{i(\mathbf{k}\cdot\mathbf{x} - \omega t)} \langle \rho(\mathbf{x}, t) \cdot \rho(0, 0) \rangle \quad \text{(Dynamic structure factor)}$$

where ρ is the mass density, and $\mathbf{J} = \rho\mathbf{v}$ is the current density, $\mathbf{v}$ being the velocity.

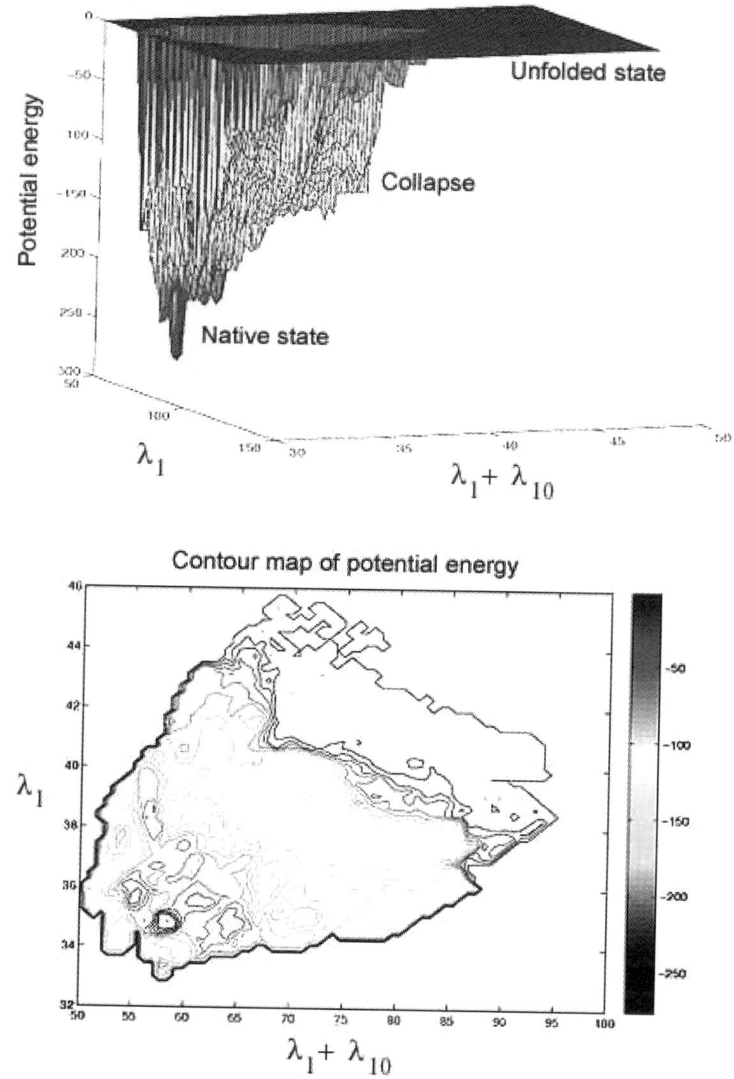

Figure 14. Potential energy surface and contour map. The axis are λ_1 and $\lambda_1 + \lambda_{10}$, where λ_n is the nth eigenvalue of the 10x10 matrix of distances between residues..

Some preliminary results by Dr. Weitao Sun of Tsinghua University are shown in Fig.18.

12. Outlook

We plan to further develop CSAW on two fronts:

1. Structural prediction and design:

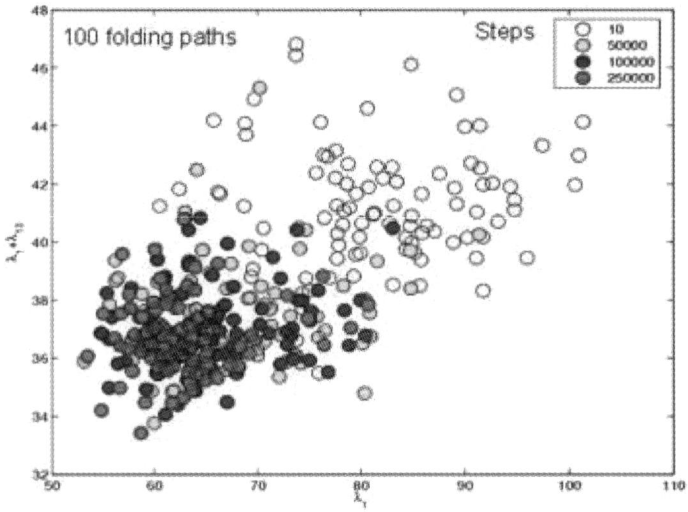

Figure 15. Folding paths, showing convergence from the unfolded states (white), througth intermediate states (black), to the native state represented by an "attractor" (gray).

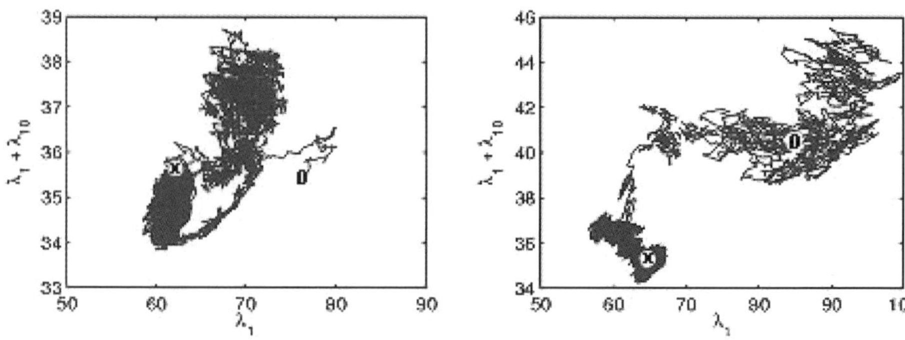

Figure 16. Samples paths, with starting point marked 0, and endpoint marked x. The state spents a long time searching for the narrow corridor leading to the "attractor".

- Accurate simulation of real proteins by CSAW
- CSAW as front end for molecular dynamics

2.Theoretical laboratory:

- Calculation of properties not directly measurable
- Testing principles and mechanisms

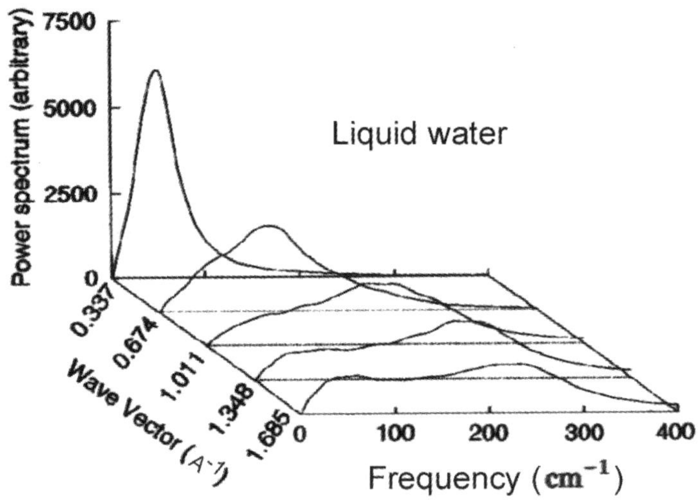

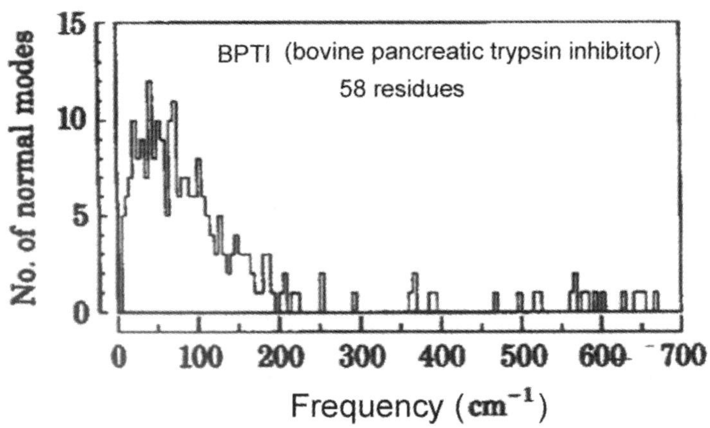

Figure 17. Samples paths, with starting point marked 0, and endpoint marked x. The state spents a long time searching for the narrow corridor leading to the "attractor".

Acknowledgments

I thank the hospitality of the Zhou Pei-Yuan Center for Applied Mathematics, a center established by Professor Lin at Tsinghua University, where I was a visitor while developing CSAW.

References

1. C.C.Lin, *Acta Mechanica Sinica*, **19,** 97 (2003).
2. K. Huang, "CSAW: Dynamical model of protein folding", arXiv cond-mat/0601244 v1 12 Jan 2006.

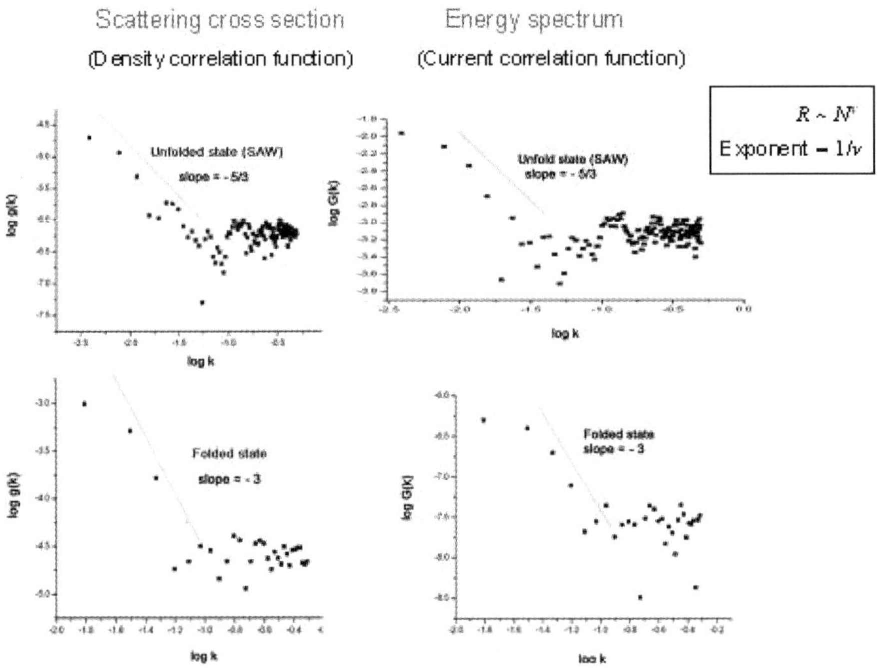

Figure 18. spectrum of a model protein obtained through correlations functions, for both the unfolded and native state. The log-log plots exhibit power-law regions. The exponents are related to the compactness index relating the radius and residue number.

3. K. Huang, *Lectures on Statistical Physics and Protein Folding* (World Scientific Publishing, Singapore, 2005).
4. B. Li, N. Madras, and A.D. Sokal, *J. Stat. Phys.* **80**, 661 (1995).
5. T. Kennedy, *J. Stat. Phys.* **106**, 407 (2002).
6. H. Li, C. Tang, N.N. Wingren, *Proc. Nat. Acad. Sci., USA*, **95**, 4987-4990 (1998).
7. J. H. Conway and N.J.A. Sloane, *Sphere Packings, Lattices, and Groups*, 2nd ed. (Springer-Verlag, New York, 1993).
8. T. Uzawa, S. Akiyama, T. Kimura, S. Takahashi, K. Ishimori, I. Morishima, T. Fujisawa, *Proc. Nat. Acad. Sci., USA*, **101**, 1171-1176 (2004).
9. J. Z. Lei and K. Huang, "CSAW II: Folding pathways of chignolin", (in preparation).

"COLLAPSE AND SEARCH" DYNAMICS OF PROTEIN FOLDING DETECTED BY TIME-RESOLVED SMALL-ANGLE X-RAY SCATTERING*

SATOSHI TAKAHASHI

Institute for Protein Research, Osaka University / CREST, JST
Suita Osaka 565-0871, Japan

TETSURO FUJISAWA

RIKEN Harima Institute SPring-8 Center, Biometal Science Laboratory
Sayo Hyogo 679-5148, Japan

The polypeptide collapse is an essential dynamics in protein folding. To understand the mechanism of the collapse, the changes in protein compactness in the process of protein folding was observed by using time-resolved small-angle x-ray scattering. The characterized dynamics for several proteins with different topologies suggested a common folding mechanism termed "collapse and search" dynamics, in which the polypeptide collapse precedes the formation of the native contact formation. The collapsed intermediates demonstrated the scaling relationship between radius of gyration (R_g) and chain length with a scaling exponent of 0.35 ± 0.11, which is close to the theoretical value (1/3) for the collapsed globules predicted by a statistical theory of polymers. Thus, it was suggested that the initial collapse is caused by the coil-globule transition of polymers. Since the collapse is essential to the folding of larger proteins, further investigations on the collapse likely lead to an important insight into the protein folding phenomena.

1 Introduction

Proteins collapse from extended conformations of unfolded polypeptides to tightly packed and folded structures. Therefore, the polypeptide collapse is an essential dynamics in the mechanism of protein folding. The importance of the collapse is reflected in various classical proposals of protein folding such as "hydrophobic collapse" and "framework" models, in which the timing of polypeptide collapse is considered to be the key structural events.[1,2] Despite the acknowledged importance of the collapse, its molecular mechanism has long been unknown due to the scarcity of experimental data on the dynamics of collapse in protein folding.

Small-angle X-ray scattering (SAXS) is a powerful method for the structural characterization of non-crystalline macromolecules, which can be applicable for the time-resolved investigation on protein folding as demonstrated in several pioneering investigations.[3,4,5] The method becomes particularly useful with the advance of the third generation synchrotron facilities. In SPring-8, a beamline specifically designed for X-ray scattering of biological macromolecules was constructed (BL45XU-SAXS), which

---

*This work was partially supported by grants from the Ministry of Education, Science, Sports and Culture, and from Japan Science and Technology Agency.

enabled the collection of scattering data for dilute solutions with short accumulation periods.[6] Furthermore, to enable a time-resolved observation of SAXS, a continuous-flow rapid mixing apparatus for the initiation of protein refolding was developed.[7,8] By combining the mixing device and the synchrotron x-ray source, we were successful in investigating protein compaction process with a time resolution of several hundreds of microseconds, which was more than 10 times of improvements in the time resolution over the previous devices.[9] The time resolution was further found to be suitable to resolve the compaction in the folding process of many proteins.

To understand the molecular mechanism of the collapse in protein folding, we have been systematically investigating the kinetic processes of protein folding by using the developed device.[9-13] The characterized dynamics for several proteins with different chain lengths and secondary structure contents suggested a common folding mechanism termed "collapse and search" dynamics, in which the initial collapse promotes the search for the correct secondary and tertiary structures. We will summarize the results and insights obtained through the efforts of our group.

2 Stepwise Folding of Cytochrome *c*

Cytochrome *c* (cyt *c*) is a small globular protein of 104 amino acids. A covalently attached heme is surrounded by three helices called N-terminal, C-terminal and 60's helices.[14] The time-resolved circular dichroism (CD) measurements clarified the stepwise formation of these helices through two folding intermediates having ~500 μs and 7 ms of the lifetimes.[8] However, the compactness of these intermediates was not evaluated.

We conducted the time-resolved SAXS measurements on the folding process of cyt *c*, which was initiated by a rapid pH jump from pH 2.2 to 4.5.[9] The acid unfolded protein (R_g = 24.3 Å) converts to the initial intermediate (intermediate I) within 160 μs, resulting in ~ 4 Å reduction of R_g. Intermediate I contains a small hydrophobic cluster and an extended region, and ~20 % of the native helical content. The conversion of intermediate I to the second intermediate (intermediate II) occurs with a time constant of ~500 μs, whose helix content (~70 % native) and R_g (R_g=17.7 Å) are very similar to those of the static molten globule state.[15] The similarity suggests that intermediate II has the structured core including the tree major helices. The native structure (R_g=13.8 Å) is formed from intermediate II upon the final collapse that takes ~15 ms.

The result on cyt *c* was the first demonstration of the significant collapse (~40 % of total change in R_g) that occurs within the dead time of the developed device. Furthermore, limited amount of the secondary structure content in the initial intermediate was also demonstrated. Since intermediate I likely possesses the collapsed core involving the specific region of the protein, we conclude that cyt *c* folding proceeds with the specific collapse followed by the cooperative acquisition of secondary structure and compactness.

3 Collapse and Search Dynamics of Apomyoglobin Folding

As the second target protein, we chose apomyoglobin (apoMb), which possesses globular shape with R_g of 18.2 Å and seven helices labeled A~E, G and H at neutral pH.[16] The protein becomes acid-unfolded state at pH 2.2 possessing a large R_g (29.7 Å), and forms a static intermediate at pH 4.2 with a significant helical content (33 %) attributed to A, G and H helices.[17] It was demonstrated that a kinetic intermediate is formed within the dead time (~5 ms) of stopped-flow devices, whose conformation resembles the static intermediate stabilized at pH 4.2.[18] However, the mechanism of the intermediate formation in the submillisecond time domain was largely unknown.

To investigate the compaction process, the folding of apoMb from the acid-unfolded state at pH 2.2 was initiated by a pH jump to 6.0 and was monitored by the time-resolved SAXS method.[10] A significant collapse, corresponding to ~50 % of the overall change in R_g from the unfolded to native conformation, was observed within 300 μs after the pH jump. The collapsed intermediate has a helical content of ~33 %. Subsequently, the second intermediate was formed with a time constant of ~5 ms, and possesses an increased helical content (~44 %) without detectable changes in R_g. Finally, the native state (R_g = 18.2Å) was formed from the second intermediate with a time constant of ~50 ms.

The first characteristic observation in the above dynamics of apoMb is the initial and significant collapse that occurs within the dead time of the experiments (<300 μs). The amount of collapsed domain roughly estimated from the scattering data was ~80 %, and cannot be explained only by the A, G and H helices (~40 %). Thus, the initial collapse likely involves regions corresponding not only the structured helices but also the unstructured helices. The second characteristic observation is the processes after the initial intermediate, which can be described as the stepwise helix formations within the collapsed conformation occurring in several to several hundreds of milliseconds. Based on these observations, we termed the folding mechanism of apoMb as "collapse and search" dynamics, in which the initial collapse precedes the search process for the native structures.

4 A Significant Collapse in the Folding of Single Chain Monellin

The above proteins investigated by the time-resolved SAXS method were α-helical. To compare the folding dynamics of proteins with different secondary structures, we next chose single chain monellin (SMN), which contains a 5-stranded antiparallel β-sheet docked with an α-helix and no disulfide bridge.[19] Since SMN demonstrates a reversible alkaline unfolding, we investigated the folding dynamics of SMN by using a pH jump from the alkaline denatured state.[11]

The burst phase in the folding of SMN occurs within 300 μs and forms the initial intermediate (I_1). The amount of secondary structure for I_1 roughly estimated from the time-resolved CD spectrum is ~6 % and is comparable to that of the initial unfolded conformation (~3 %). In contrast, the SAXS data demonstrated that I_1 possesses a collapsed R_g (18.2 Å), which is significantly smaller than that of the initial unfolded conformation (R_g = 24.3 Å). Thus, the burst phase can be described as a significant collapse with a small amount of secondary structure formation. The second intermediate (I_2) is formed with a time constant of ~14 ms and possesses a further collapsed conformation whose overall shape is very similar to that of the native. The secondary structure content of I_2 is ~22 % and is still smaller than that of the native state (~52 %). The final phase, corresponding to the transition from I_2 to the native state (R_g = 15.8 Å), occurs with the time constant of ~1.3 s. Since the large amplitude of the CD signal develops in this phase, the formation of main chain H-bonds characterizes the final phase.

The remarkable differences in the folding dynamics of α-helical and β-sheet proteins can be recognized in the folding trajectories of SMN, apoMb and cyt *c* presented in the two-dimensional space defined by R_g and the amount of secondary structures (Figure 1). The "L-shaped" trajectory of SMN contrasts with the more diagonal trajectories of helical proteins, indicating that the folding landscapes are distinct for proteins composed of different secondary structures. Brooks III *et al.* found for the helical proteins that tertiary and secondary structures are formed more or less cooperatively.[20] In contrast, the calculations on β-sheet proteins showed the L-shaped surface. Our results constitute the clear experimental verification of the rapid collapse with a small increase in the secondary structure content for a β-sheet protein.

In addition to the differences, we can point out the remarkable similarity in the folding dynamics of the three proteins with different secondary structure contents. Namely, the significant collapse of polypeptides was commonly observed as the initial dynamics occurring within the deadtime of our device. We therefore proposed that the "collapse and search" mechanism is likely the common folding mechanism of proteins.

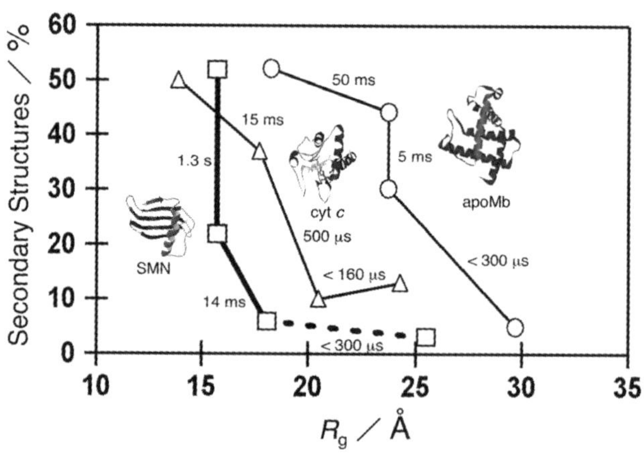

Figure 1. The folding trajectories of cyt c (triangles), apoMb (circles) and SMN (rectangles) in the two-dimensional space defined by secondary structure content and R_g. Modified from Figure 4 of ref 11.

5 Coil-Globule Transition of the Initial Collapse

The observation that the folding of proteins with different topologies commonly demonstrated the initial and significant collapse suggested that the collapse is caused not by the individual properties of proteins such as the primary sequences but by the physical properties of proteins as polymeric molecules. As the next research subject, we intended to explain the initial collapse based on Flory's theory of polymers.

The basic theory of polymer conformations was developed by Flory, in which the phase transition of polymers from an extended coil to a collapsed globule was explained and termed as the coil-globule transition.[21] The transition appeared to correspond to the collapse process of proteins; however, no clear evidence for the suggestion was presented. Based on Flory's theory a polymer, in the absence of short-distance interaction, collapses to a globule at temperatures below θ temperature (T_θ), whose R_g is described by the scaling law:

$$R_g = a\,N^{\nu} \qquad (1)$$

where N is the number of monomers, a is a parameter and ν is a scaling exponent and is 1/3. In contrast, at above T_θ the polymer expands and possesses a scaling exponent close to 3/5. Denatured proteins in the presence of high concentrations of denaturants are known to obey the scaling relationship with the scaling exponent close to 3/5, and were classified as the coils.[22] To assign that the collapsed intermediates are the globules, the scaling relationship and the exponent for the intermediates should be established.

To examine the scaling behavior of the initially collapsed intermediates, it is necessary to investigate proteins with various chain lengths. Thus, we carefully investigated the folding process of HO, which is the longest protein (263 residues) ever characterized by the time-resolved SAXS method.[13] HO is a highly helical protein having neither S-S disulfide bridges nor prosthetic groups; however, demonstrates a complicated folding mechanism due to a cis-trans isomerization of X-proline peptide bonds and to an oligomer formation. We resolved the complications by carefully conducting various double jump experiments and by examining concentration dependencies. We could obtain the R_g value for the initially collapsed and monomeric conformation of HO.

The R_g values for HO and other proteins immediately after the folding reaction were plotted against chain lengths in Figure 2. The values for the seven proteins with more than ~100 residues are well fitted by the scaling law with ν = 0.35 ± 0.11, which is close to 1/3 predicted for the globules. The observation suggests that the collapsed intermediates correspond to the globules in poor solvent explained by Flory's theory.

We omitted three small proteins from the above scaling law; however, propose that the omission might be explained by the chain-length dependency of the coil-globule transition. It was observed that proteins of less than 100 residues usually do not accumulate kinetic intermediates, and possess the expanded R_g after the refolding jump.[5,23] We propose that T_θ for small proteins is lower than the experimental temperatures causing small proteins to remain expanded until the moment of folding. It is also possible that the expanded conformation might be the result of reduced cooperativity of the coil-globule transition whose width is proportional to $N^{-1/2}$ and profoundly increases for flexible chain.[24] Thus, the coil-globule transition offers a general explanation for the initial folding mechanisms of small and larger proteins.

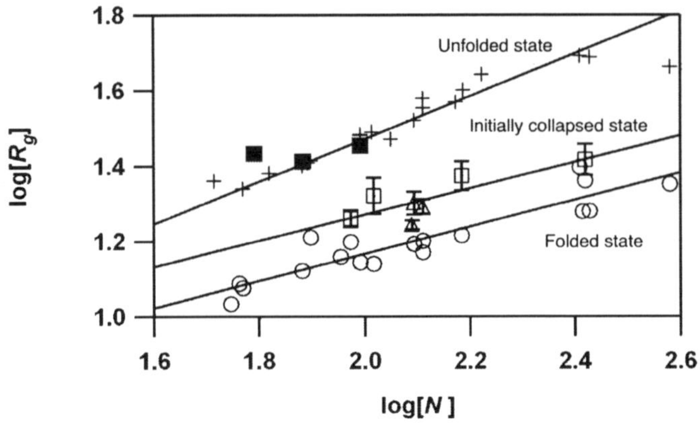

Figure 2. The correlation plot between R_g and N for the chemically unfolded state (crosses), initially collapsed state (open squares and rectangles) and native state (circles). The R_g values for the small proteins that are expanded after the refolding reaction are shown as filled squares.[5,23] Modified from Figure 5 of ref 13.

6 Summary and Perspective

In the series of investigation on the dynamics of protein folding, we observed that the folding transitions of proteins with more than 100 residues generally follow the collapse and search mechanism, in which the rapid collapse and the subsequent search in the collapsed conformations are the characteristic events. Further, we presented evidence suggesting that in spite of large diversity of proteins the initial collapse is caused by a very simplified scheme, *i.e.*, the coil-globule transition. We stress that the results could be obtained only by the *in situ* time-resolved SAXS method developed through the efforts of our group. The method should be applicable for the understanding of other biological processes, in which rapid conformational changes are expected.

One of the major purposes in protein folding research is the prediction of folded structure of proteins from the primary sequences of amino acid residues. Despite the significant advances in the recent prediction methods, it is known that the prediction for proteins that are larger than 100 residues is still extremely difficult.[25] A further understanding on the molecular mechanism of the collapse will likely give an important insight that might help improve the structure prediction for larger proteins, since the difference in the folding mechanisms between proteins smaller and larger than 100 residues, resides in the absence and presence of the initial collapse event. Improvements in the time-resolution and the quality of the SAXS data for the characterization of the chain collapse should reveal rich structural events involved in the process.

Acknowledgments

We thank Drs. Shuji Akiyama, Tetsunari Kimura and Takanori Uzawa for their devotion and excitements to the project, and other collaborators for their supports and encouragements.

References

1. K. A. Dill, *Biochemistry* **24**, 1501-1509 (1985).
2. P. S. Kim and R. L. Baldwin, *Ann. Rev. Biochem.* **51**, 459-489 (1982).
3. D. Eliezer, P. A. Jennings, P. E. Wright, S. Doniach, K. O. Hodgson and H. Tsuruta, *Science* **270**, 487-488 (1995).
4. M. Arai, T. Ikura, G. V. Semisotnov, H. Kihara, Y. Amemiya, and K. Kuwajima, *J. Mol. Biol.* **275**, 149-162 (1998).
5. K. W. Plaxco, I. S. Millet, D. J. Segel, S. Doniach and D. Baker, *Nature Struct. Biol.* **6**, 554-556 (1999).
6. T. Fujisawa, K. Inoue, T. Oka, H. Iwamoto, T. Uruga, T. Kumasaka, Y, Inoko, N. Yagi, M. Yamamoto and T. Ueki, *J. Appl. Crystallogr.* **33**, 797-800 (2000).
7. S. Takahashi, S.-R. Yeh, T. K. Das, C.-K. Chan, D. S. Gottfried and D. L. Rousseau, *Nature, Struct. Biol.* **4**, 44-50 (1997).
8. S. Akiyama, S. Takahashi, K. Ishimori and I. Morishima, *Nature Struct. Biol.* **7**, 514-520 (2000).
9. S. Akiyama, S. Takahashi, T. Kimura, K. Ishimori, I. Morishima, Y. Nishikawa and T. Fujisawa, *Proc. Natl. Acad. Sci. USA* **99**, 1329-1334 (2002).
10. T. Uzawa, S. Akiyama, T. Kimura, S. Takahashi, K. Ishimori, I. Morishima and T. Fujisawa, *Proc. Natl. Acad. Sci. USA* **101**, 1171-1176 (2004).
11. T. Kimura, T. Uzawa, K. Ishimori, I. Morishima, S. Takahashi, T. Konno, S. Akiyama and T. Fujisawa, *Proc. Natl. Acad. Sci. USA* **102**, 2748-2753 (2005).
12. T. Kimura, S. Akiyama, T. Uzawa, K. Ishimori, I. Morishima, T. Fujisawa and S. Takahashi, *J. Mol. Biol.* **350**, 349-362 (2005).
13. T. Uzawa, T. Kimura, K. Ishimori, I. Morishima, T. Matsui, M. Ikeda-Saito, S. Takahashi, S. Akiyama and T. Fujisawa, *J. Mol. Biol.* **357**, 997-1008 (2006).
14. G. W. Bushnell, G. V. Louie and G. D. Brayer, *J. Mol. Biol.* **214**, 585-595.
15. M. Kataoka, Y. Hagihara, K. Mihara and Y. Goto, *J. Mol. Biol.* **229**, 591-596 (1993).
16. D. Eliezer and P. E. Wright, *J. Mol. Biol.* **263**, 531-538 (1996).
17. F. M. Hughson, P. E. Wright and R. L. Baldwin, *Science* **249**, 1544-1548 (1990).
18. P. A. Jennings and W. E. Wright, *Science* **262**, 892-896 (1993).

19. S. Y. Lee, J. H. Lee, H. J. Chang, J. M. Cho, J. W. Jung and W. T. Lee, *Biochemistry* **38**, 2340-2346 (1999).
20. C. L. Brooks III, *Acc. Chem. Res.* **35**, 447-454 (2002).
21. P. J. Flory, *Principles of Polymer Chemistry*, Cornel University Press, Ithaca, NY (1953).
22. J. E. Kohn, I. S. Millett, J. Jacob, B. Zagrovic, T. M. Dillon, N. Cingel, R. S. Dothager, S. Seifert, P. Thiyagarajan, T. R. Sosnick, M. Z. Hasan, V. S. Pande, I. Ruczinski, S. Doniach and K. W. Plaxco, *Proc. Natl. Acad. Sci. USA* **101**, 12491-12496 (2004).
23. J. Jacob, B. Krantz, R. S. Dothager, P. Thiyagarajan and T. R. Sosnick, *J. Mol. Biol.* **277**, 985-994 (2004).
24. A. Grosberg and A. Khokhlov, *Statistical Physics of Macromolecules*, American Institute of Physics, New York (1994).
25. J. Moult, *Curr. Opin. Struct. Biol.* **15**, 285-289 (2005).

STRUCTURAL TRANSITIONS IN BIOPOLYMERS: FROM DNA TO PROTEIN TO SPIDER SILK

HAIJUN ZHOU*
Institute of Theoretical Physics, the Chinese Academy of Sciences, Beijing 100080, China
**E-mail: zhouhj@itp.ac.cn*

Biopolymers are ubiquitous in a biological living system. They are involved in almost all cellular biochemical reactions, playing versatile biological roles. The mechanical characteristic of a biopolymer is directly related to its functions. In this paper, we briefly review theoretical efforts of our group on the structural transitions of three major types of biopolymers (DNA, RNA, proteins) and on the structural organization in the spider capture silk. These theoretical studies demonstrated that, weak non-covalent interactions, such as wan der Waals attraction, hydrogen bonding, hydrophobic interaction, and screened electrostatic interactions, enable a biopolymer to achieve an optimal combination of structural stability and deformability in accordance with its specific biological functions.

1. Introduction

At physiological conditions, the typical force involved in a biological process is of the order of piconewtons (1 pN $= 10^{-12}$ N). Such high levels of force resolution have been achieved in the last two decades by using singe-molecule manipulation methods. For example, a magnetic tweezer can generate a constant force in the range 0.01 pN – 100 pN, an optical tweezer can generate forces 0.1 pN – 150 pN, and an atomic force microscopy can generate forces 5 pN – 10 nN (1 nN $= 10^{-9}$ N $= 10^3$ pN). With these new instruments, the response of a single biopolymer under the action of an external force can be measured and structural transitions in a biopolymer can therefore be detected. The first biopolymer to be studied this way is the double–stranded DNA (dsDNA), the genetic material of a biological organism[1]. Later on, the mechanical properties of single–stranded DNA (ssDNA), RNA, and protein molecules such as titin and silk proteins were also studied. These were followed by single-molecular studies on the interactions between DNA and various proteins, including DNA polymerases, DNA topoisomerases, helicases, and other motors/enzymes. The structural organizations of eukaryotic chromosomes were also probed by mechanical experiments.

Biopolymers DNA, RNAs, and proteins are the most important macromolecules in a living cell. DNA stores the genetic information of a cell. This information is translated into protein sequences by way of RNAs. After a protein sequence is folded into its native three-dimensional structure, it is then ready to perform its biological roles (such as gene-expression regulation, signal transduction, etc.). There are many very complicated physical interactions among DNA, RNAs, and proteins. The mechanical property of a biopolymer is therefore directly related to its functions. The accumulation of single-molecular force manipulation data make it possible to gain quantitative understanding of the mechanical properties of these important biopolymers. In the last several years, we have performed a series of theoretical investigations on the structural transitions of DNA, RNAs, and proteins. We have also studied the elasticity of spider capture silk. These theoretical studies demonstrated that, weak non-covalent interactions, such as wan der Waals attraction, hydrogen bonding, hydrophobic interaction, and screened electrostatic interactions, enable a biopolymer to achieve an optimal combination of structural stability and deformability in accordance with its specific biological functions.

This paper is a brief review of our own theoretical efforts. The review article Ref. 2 contains an expanded overview of the theoretical progresses on single-biopolymer modeling. For readers who wishes to get an overview of the experimental development, the review articles Refs. 3, 4, 5, 6, 7 can be consulted.

This review paper is organized as follows. We first mention three simple models for biopolymers in the next section. In Sec. 3 the mechanical property of dsDNA is studied; followed by the RNA and ssDNA secondary structure denaturation in Sec. 4. Section 5 investigate the unfolding process of protein β-sheet secondary structures. In Sec. 6 we present a hierarchical chain model to account for the exponential force-extension relationship of spider capture silk. The last section concludes this paper.

2. Simple polymer elasticity models

As a starting point, we list in this section the major properties of three simple polymer models. Related information on polymer elasticity models can be found in text books such as Ref. 8 and Ref. 9. Due to sequence heterogeneity and complicated intra-polymer and inter-polymer interactions, real biopolymers are of cause more complex than these simple models. By comparing the behavior of a real biopolymer with those of the model polymers, we may be able to distinguish mechanical properties that are specific to this biopolymer from mechanical properties that are general to all polymers.

2.1. *The Gaussian chain model*

The simplest polymer model is the Gaussian chain model. This model regards a polymer as composed of N beads. At time point t, each bead i is in position $\mathbf{r_i(t)}$; between two consecutive beads $i-1$ and i there is the following harmonic interaction

$$E_{i-1,i} = \frac{3}{2}\frac{k_B T}{b^2}(\mathbf{r}_i - \mathbf{r}_{i-1})^2 \ , \tag{1}$$

where k_B is Boltzmann's constant, T is the temperature, and b is the Kuhn length. The Kuhn length b characterizes the mean square root distance between bead $i-1$

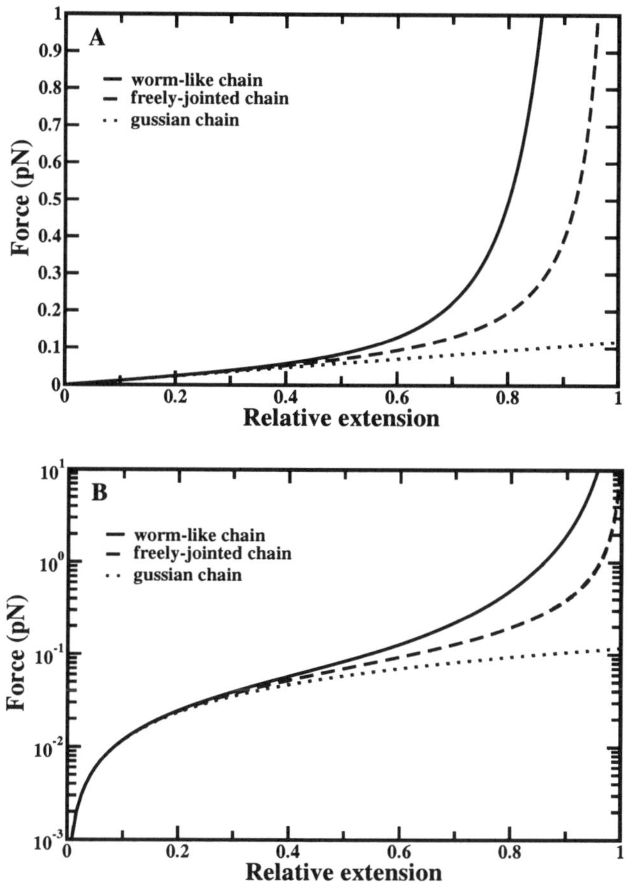

Figure 1. Force-extension relations for a Gaussian chain (dotted line), a freely-jointed chain (dashed line), and a wormlike-chain (solid line). All three curves are produced with the same Kuhn length $b = 100$ nm and the same temperature $T = 300$ K. In (A), force is in linear scale while in (B), force is in logarithmic scale.

and bead i in the absence of external force. Under the action of an external force of magnitude f, the extension z of the Gaussian chain along the force direction is

$$z = \frac{Nb}{3} \frac{bf}{k_B T} \,. \tag{2}$$

This is a *linear* force-extension relation (see the dotted lines of Fig. 1). From Eq. (2), one know that the characteristic length scale of the Gaussian chain is b, and the characteristic force scale of the Gaussian chain is $k_B T/b$. For a biopolymer such as RNA and proteins, the Kuhn length b is in the order of nanometer (1 nm $= 10^{-9}$ m). Since the temperature $T \approx 300$ K, we obtain that the characteristic force is ≈ 4 pN.

Although the Gaussian chain model can describe the response of a single polymer chain under the stretching of a small low force, it is obviously an over-simplified model. One sever simplification is that the distance between two adjacent beads can be extended without bound. The freely jointed chain (FJC) model improves on this point.

2.2. *The freely jointed chain (FJC) model*

In the FJC model, the polymer is composed of a tandem sequence of N rigid rods each of contour length b. Each rod can take any orientation irrespective of the orientations of other rods. Under the action of an external force, these rigid rods will align along the force direction, and the total extension z along the force direction reads

$$z = NbL\big(fb/k_BT\big) \,, \tag{3}$$

where $L(x)$ is the Langevin function defined by

$$L(x) = 1/\tanh x - 1/x \,. \tag{4}$$

In the limit of $f \ll k_BT/b$, Eq. (3) reduces to Eq. (2) of the Gaussian chain model. On the other hand, when $f \gg k_BT/b$, Eq. (3) can be re-written in the form

$$z = Nb\left(1 - \frac{k_BT}{fb}\right) ; \tag{5}$$

that is, the difference between the total contour length of the polymer and the total extension, $Nb - z$, scales with the external force with an *inverse linear* law (dashed lines of Fig. 1).

The FJC model can be extended to take into account the finite extensibility of each rod. This is achieved via the introduction of a stretching modulus. The extensible FJC model (eFJC) is able to describe the mechanical response of an extended single-stranded DNA (ssDNA) or protein chain fairly good when the external force is of the order of 10 pN.

The FJC model also have an important simplification, namely, the bending energy contribution of the polymer chain is not considered. This shortcoming is overcome by the worm-like chain model of the next subsection.

2.3. *The worm-like chain (WLC) model*

The WLC model regards a polymer as an inextensible continuous string of total contour length L. The tangential vector of the string at contour length s is denoted as $\hat{\mathbf{t}}(s)$. The coordinate of the string at contour length length s is $\mathbf{r}(s) = \mathbf{r}(0) + \int_0^s \hat{\mathbf{t}}(s')\mathrm{d}s'$. For each configuration of the string there is an associated bending energy

$$E[\{\hat{\mathbf{t}}(s)\}] = \int_0^L \frac{k_BTb}{4}\left(\frac{\mathrm{d}\hat{\mathbf{t}}(s)}{\mathrm{d}s}\right)^2 \mathrm{d}s \,. \tag{6}$$

At the action of an external force of magnitude f, the extension z of the string along the force direction can be obtained numerically by transfer matrix method [10]. There also exists an excellent variational expression of z as a function of f. This variational expression reads [10]

$$z = LL(\eta) , \tag{7}$$

where η is determined by the external force f by the following equation

$$\frac{2fb}{k_B T} = \eta + \frac{L(\eta)}{L'(\eta)} , \tag{8}$$

with $L'(\eta) = \mathrm{d}L(\eta)/\mathrm{d}\eta$. Furthermore, a simple interpolation formula can also be written down for the WLC model [11,10]:

$$f = \frac{k_B T}{2b}\Big(\frac{1}{(1-z/L)^2} - 1 + 4z/L\Big) . \tag{9}$$

When the external force $f \gg k_B T/b$, the force–extension relationship of the WLC model reduces to Eq. (2) of the Gaussian chain model. On the other hand, when $f \gg k_B T/b$, it can be shown [10] that

$$z = L\Big(1 - \sqrt{\frac{k_B T}{2fb}}\Big) ; \tag{10}$$

in other words, the difference between the extension z and the total contour length L is proportional to $f^{-1/2}$ for large forces (solid lines of Fig. 1). This scaling behavior characterizes the WLC chain model. It was first revealed by Bustamante and co-workers [11] that, the WLC model is an excellent model for double-stranded DNA at the force regime of $f \leq 10$ pN. The worm-like chain model with $b \approx 1$ nm can also fit the force-extension curves of extended ssDNA and proteins very well.

What is the physical meaning of the length scale b in Eq. (6)? For the WLC model, usually another quantity, called the bending persistence length ℓ_b, is introduced. For three-dimensional worm-like polymers, the bending persistence length is related to the Kuhn length b by the simple relation that $\ell_b = b/2$. The physical meaning of ℓ_b is as follows: Suppose that the tangential vector of the string is $\hat{\mathbf{t}}_0$ at arc length s, then the tangential vector $\hat{\mathbf{t}}_1$ at arc length s' is correlated to $\hat{\mathbf{t}}_0$ through

$$\langle \hat{\mathbf{t}}_1 \cdot \hat{\mathbf{t}}_0 \rangle = \exp\Big(-\frac{|s'-s|}{\ell_p}\Big) . \tag{11}$$

The bending persistence length ℓ_b (or equivalently the Kuhn length b) is the orientation correlation length of a semiflexible polymer.

2.4. *Comparison of the three polymer models*

As a summary of this section, we demonstrate the force–extension curves predicted by the above–mentioned three polymer elasticity models in Fig. 1 for the case of $b =$

100 nm and $T = 300$ K. When the external force $f < 0.04$ pN, all three theoretical curves superimpose onto each other. The differences between these models become significant only when $f \gg 0.04$ pN.

3. The over-stretching transition of double-stranded DNA

It is now widely accepted that, under low or moderate external stretching (force lower than 10 pN), dsDNA can be regarded as an inextensible semiflexible polymer. At physiological salt conditions, the mechanical response can be well approximated by a worm-like chain of bending persistence $\ell_b \simeq 53$ nm [1,11,10]. This is the entropic elasticity regime. In this regime, the external force tends to pull the dsDNA straight, while the thermal noise due to the collision between dsDNA and solution molecules tend to make dsDNA to be in a coiled and curved form. The equilibrium reached between the competition of these two effects determines the total extension of the polymer chain. On the other hand, the dsDNA double helix is not affected by external stretching.

The bending persistence length has two sources of contributions: from the local steric effect between adjacent DNA base–pairs; from the electrostatic repulsive interactions, since dsDNA is a negatively charged polyelectrolyte. Changing the solution salt conditions has an effect on dsDNA bending persistence length.

When the external force is further increased ($f \geq 10$ pN), dsDNA is more than 95% aligned along the force direction. In response to the external stretching, therefore dsDNA will be stretched by increasing slightly the vertical distance between two adjacent base–pairs. This leads to a stretching modulus of the order of 1000 pN, a value that is consistent with the bending persistence length of dsDNA [12,3].

At force $f \approx 70$ pN, a new phenomenon was observed [13,12]. The total extension of a dsDNA polymer changes from its B-form contour length to about 1.7 times its B-form contour length over a narrow force range of about 5 pN. Furthermore, this over-stretching transition is almost reversible, indicating that it is an equilibrium process. The over-stretched DNA is referred to as the S-form DNA in the literature. The nature of the DNA over-stretching transition is extensively investigated by many groups. The most important physical reason for this highly cooperative over-stretching transition may be the weak base–pair stacking interaction of dsDNA [14].

To quantitatively understand the dsDNA over-stretching transition, Zhou and co-authors [14,15] introduced a semi-microscopic model for dsDNA. This double–stranded polymer model has the following ingredients: (1) the polymer is composed of two inextensible single–stranded chain; (2) these two strands are bound into a double–strand through many lateral base–pairs; (3) between two adjacent base–pairs there is a vertical base–pair stacking interaction. The base-pair stacking potential has the following Lennard-Jones form

$$U_{\text{stack}}(r_{i,i+1}) = \begin{cases} \epsilon_0 \left[\left(\frac{r_{i,i+1}}{r_0}\right)^{12} - 2\left(\frac{r_{i,i+1}}{r_0}\right)^{6} \right] & \text{if right-hand stacking}, \\ \text{constant} & \text{if left-hand stacking}. \end{cases} \tag{12}$$

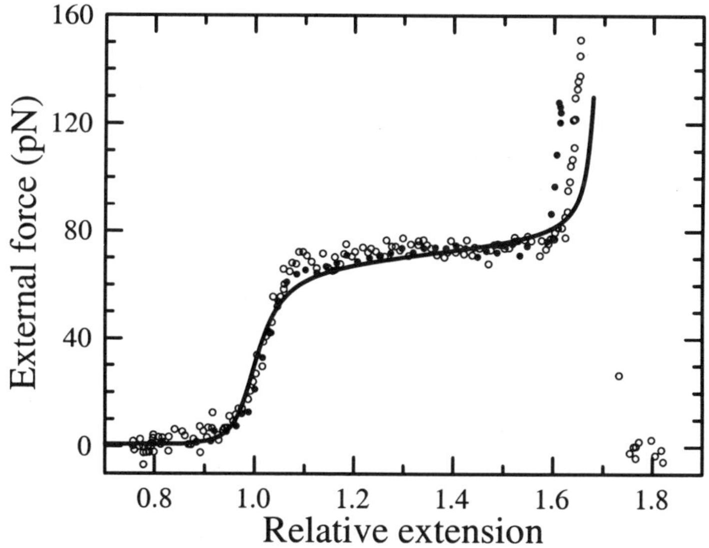

Figure 2. Force-extension response curve for a double-stranded DNA. The open and closed circles are two sets of experimental measurements from Ref. 13; while the solid line is the result of the theoretical model of Ref. 14. The length unit of the horizontal axis is the total contour length of the DNA in its relaxed B-form conformation.

In Eq. (12), ϵ_0 is the base-pair stacking energy constant; r_0 is the optimal vertical distance between two base-pairs, $r_{i,i+1}$ is the actual vertical distance between base-pairs i and $i+1$. The stacking potential also depends on whether the two adjacent base-pairs are right-handedly stacked or left-handedly stacked. This later consideration ensures that dsDNA in its relaxed state will be a right-handed double-helix.

Zhou and co-authors were able to solve the above-mentioned double-stranded polymer model exactly by using Green function method. The details of their model and their analytical calculations are documented in Ref. 15, here we only mention the main predictions of their theoretical model.

Figure 2 shows the experimental and theoretical force–extension response curves for a long dsDNA. The agreement between the experimental observations of Ref. 13 and the theoretical prediction of the theoretical model is satisfactory, since the theoretical curve of Fig. 2 was obtained with only one fitting parameter. From the viewpoint of base-pair stacking interaction, the over-stretching transition is understood as follows: First, the base–pair stacking interaction is relatively strong, of the order of 10 k_BT at room temperature. Therefore when the external force is low or moderate, the dsDNA base–pairs are tightly stacked onto each other; this explains why the stretching modulus of dsDNA is very large. On the other hand, the base–pair stacking interaction is a weak interaction, it has only finite interaction range, beyond that range the interaction drops quickly. Therefore, when the inter–base–pair distance increase to a certain critical level, the base–pair stacking interaction can no longer withstand the external force; consequently, the double–helix is untwisted and form a ladder-like structure.

In real situations, the over-stretching transition may be much more complex than was assumed in the above-mentioned model. For example, in the over-stretched S-form, the base–pairs may be tilted with respect to the central axis to gain some residual base–pair stacking. It may also happen that the two strands will be separated from each other by breakage of base–pair hydrogen bonds [5]. In a complete model of dsDNA over-stretching, all these effects should be included.

Since dsDNA is a double-stranded polymer, an important form of deformation is twisting of the double-strand. This twisting causes local torsion in the dsDNA chain. To relax this twisting stress, the conformation of the central line of the dsDNA chain may deform into a curved form, called DNA supercoiling. DNA supercoiling is a very significant biological phenomenon. The model described in this section can also partially explain the behavior of a highly extended and supercoiled dsDNA [14,15]. The numerical work of Zhou and Ou-Yang [16] also suggested the possibility of a left-handed double-helical DNA structure which is stabilized by external stretching and negative supercoiling.

4. Denaturation of RNA secondary structures

RNA is different from dsDNA in that it has only one poly-nucleotide strand. The structure of a RNA chain, on the other hand, can be very complicated, since nucleotide bases along the same chain can form many intra-chain base-pairs. In the absence of external stretching, a RNA chain will fold back at various points to facilitate the formation of base-pairs. The relative positioning of any two base-pairs in a RNA structure can be grouped into three types as shown in Fig. 3: independent, nested, or crossed. If one removes all the crossed base-pairs from a RNA structure in the most economic way (i.e., trying to keep as many base-pairs as possible), the remaining base-pairing pattern is referred to as a secondary structure of the RNA polymer. In most cases, the configurational energy of a RNA structure comes mainly from the configurational energy of its secondary structure. Those crossed base-pairs in a RNA structure further stabilizes the structure; however their energy contribution to the total configurational energy is not comparable to that of the secondary structure. Because of this separation of energy scales, theoretical studies on RNA usually focus on the secondary structures of RNA.

When one end of a RNA polymer is fixed and the other end is pulled with an external force, the structure of the RNA will re-organize so as to make the polymer chain more aligned along the force direction. Experimental observations [17,3,18] revealed that the force-extension response behavior of a RNA polymer is dependent on the RNA sequence. In some experiments [3,18] a naturally occurring RNA polymer was used for stretching, and the experimental data are consistent with the prediction of a force-induced second-order RNA globule-coil phase transition [19]. This globule-coil transition occurs at force $f \approx 1.0$ pN, beyond which the relative extension of the

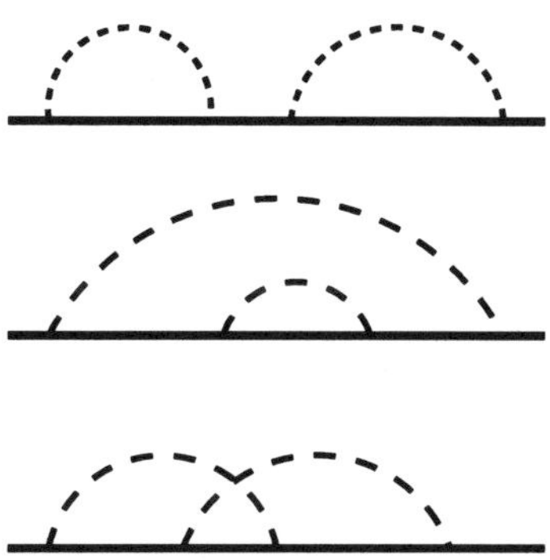

Figure 3. The relative positioning of two base-pairs along a RNA chain: (a) mutual independence; (b) nested; and (c) crossed. Thick line denotes the RNA chain and curved dashed lines denotes the formation of base-pairs.

polymer gradually increases from zero as a function of force [3,18]. On the other hand, Rief and co-authors [17] performed a RNA stretching experiment using poly-(G-C) or poly-(A-T) nucleotide chains. They found a force-plateau in the force-extension response curve. Such a force-plateau indicates that the denaturation transition in the RNA polymer is a highly cooperative process.

Zhou and co-workers performed a theoretical study [20,21,22] on RNA secondary-structure denature, with the aim of understanding the above-mentioned experiments from the same point of view. The major ingredient of their model [21] is as follows: (1) The RNA is modeled as a linear chain of N beads, with two consecutive beads along the chain being connected by an extensible bond of relaxed length b and certain stretching modulus (the eFJC model of Sec. 2.2). (2) There is short-ranged base-pairing interaction between two beads i and j if the distance between these two beads is smaller than certain value a_0 (with $a_0 \ll b$) and if both i and j have not yet formed base-pairs with other beads. (3) There is vertical base-pair stacking interaction between two adjacent base-pairs (i, j) and $(i+1, j-1)$. (4) External stretching energy. The statistical mechanical property of this self-interacting polymer was explored by analytical calculations [21] and also by Monte Carlo simulations [22]. Here we describe the main conclusions of their work (see Fig. 4).

When the mean base-pair stacking potential is negligible, the RNA globule-coil transition is determined mainly by the average strength of the base-pairing interaction. The transition is a *continuous* phase-transition in the thermodynamic limit [19]. On the other hand, when the mean base-pair stacking potential is increased, the denaturation process of RNA becomes more and more co-operative (see Fig. 4), and a force-plateau will appear in the force-extension response curve. The work of Zhou and co-authors [21] demonstrated that the nearest-neighbor base-pair stacking potential has a dramatic effect in controlling the cooperativity of the denaturation of RNA.

The mean base-pair stacking potential can be controlled by sequence. If the

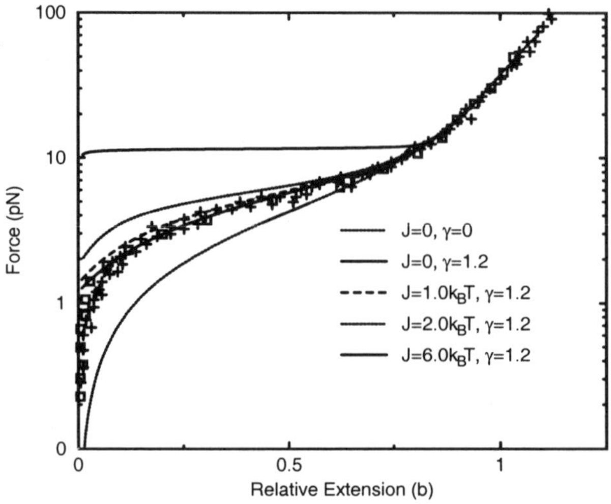

Figure 4. Force-induced RNA secondary-structure denature. The parameter γ is related to base-pairing interaction, while the parameter J is the mean base-pair stacking strength [21]. Symbols are experimental observations of Refs. 3, 18, while lines are theoretical calculations.

RNA sequence is relatively random, the probability to form two consecutive base-pairs (i, j) and $(i+1, j-1)$ is relatively low, and therefore the mean base-pair stacking potential is low; on the other hand, if the RNA chain is highly designed, then base-pair stacking possibility will be also very high.

5. The β-sheet secondary structure of proteins

The protein folding problem is one of the major issues in biophysics. At the level of secondary structures, there are α-helices and β-sheets. It has been known for more than forty years that the α-helix–coil transition is not a real phase transition in the thermodynamic limit [23]. However, the nature of the β-sheet–coil transition is not yet completely settled. Working on a simplified flexible lattice-polymer, References 24, 25, 26 all predicted that the β-sheet–coil transition is a *second-order* phase transition in the thermodynamic limit. In this section we investigate a similar 2D partially directed lattice model of protein β-sheet denaturation. We find that the order of the β-sheet–coil transition is controlled by the bending stiffness Δ of the model polymer: it is second-order for a polymer with exactly zero bending stiffness $(\Delta \equiv 0)$, while it changes to be *first-order* when the bending stiffness Δ is non-zero.

5.1. *The model and the major analytical procedure*

We study a 2D partially directed lattice polymer model as shown in Fig. 5. A chain of N identical units is located on a square lattice. The length of the bond connecting two consecutive monomers i and $i+1$ is fixed to a_0, while the direction of bond in $-\mathbf{z}_0$ is prohibited. If any two monomers i and $i+m$ $(m \geq 3)$ occupy nearest

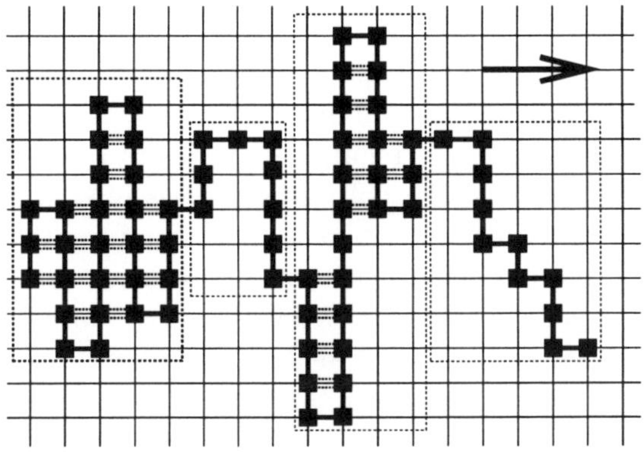

Figure 5. An 2D lattice polymer model with contacting interaction, bending stiffness, and external stretching. The arrow shows the $\mathbf{z}_0$ direction.

neighboring lattice sites, an attractive energy of magnitude ϵ is gained. Besides this 'hydrogen-bonding' interaction, we also consider the bending energy of the polymer chain. An energy penalty of magnitude Δ is added to each local direction change of the chain [27].

We follow Lifson's method [28] to calculate the free energy density of the polymer. For this purpose, a given configuration of the 2D chain is divided into a linear sequence of β-sheet segments and coil segments. A β-sheet segment is defined as a folded segment of $n_\beta \geq 2$ consecutive columns, in which contacting interactions exist between any two adjacent ones. Two consecutive β-sheet segments are separated by a coil segment, which is a segment of $n_c \geq 0$ consecutive columns in which all monomers are free of contacts. For example, the configuration shown in Fig. 5 has two β-sheets and two coils.

Under the action of an external force f, the energy of a β-sheet of n_β columns is

$$E_\beta = -\epsilon \sum_{j=1}^{n_\beta - 1} \mathrm{v}(l_j, l_{j+1}) - n_\beta f a_0 + 2(n_\beta - 1)\Delta \ , \tag{13}$$

where l_j is the number of monomers in the j-th column of the β-sheet, and $\mathrm{v}(l_j, l_{j+1}) = \min(l_j, l_{j+1}) - 1$. On the other hand, the configurational energy of a coil segment is

$$E_c = -n_c f a_0 + m_c \Delta \ , \tag{14}$$

where n_c and m_c are, respectively, the total number of columns and the total number of bends in the configuration.

To calculate the total partition function of the model system, we proceed by first calculating the partition functions of β-sheets and coils separately, and then combining them to obtain the free energy density of the whole polymer. The calculation details will be reported elsewhere [29], here we focus on the final results.

5.2. *Analytical and simulational results*

Qualitatively speaking, the β-sheet–coil transition is caused by the competition between monomer-monomer attraction and configurational entropy: Formation of contacts lowers the energy, but it requires monomers to be aligned and close to each other, thereby decreasing the polymer's degrees of configurational freedom. When the monomer-monomer attraction is much stronger than thermal energy, the polymer takes β-sheeted conformations to maximize the number of contacts. At the other limit of high temperatures, the polymer is in completely disordered coil states with maximal entropy. The analytical work of Zhou and co-authors [29] have confirmed this picture. It has been predicted that the β-sheet–coil structural transition is a real phase transition process in the thermodynamic limit. The order of this phase transition depends on whether the bending stiffness of the polymer is flexible (with $\Delta = 0$) or semi-flexible (with $\Delta > 0$).

In the case of $\Delta = 0$, the β-sheet–coil transition is *second-order*. The relative extension versus temperature curve at zero temperature is shown in Fig. 6a, and the relative extension versus force curve at constant temperature $T = 0.591\epsilon$ is shown in Fig. 6b. When there is no external force, a second-order globule–coil phase transition occurs at $T_{\rm gc}(0) = 0.8205\epsilon$. The force-induced globule–coil transition at constant temperature is also second-order. These results are in accordance with Refs. 30, 26.

In the case of positive bending stiffness, the β-sheet–coil transition becomes a first-order phase transition process. For example at $\Delta = 0.5\epsilon$ and $f = 0$, the relative extension jumps from zero to 0.193 at the transition temperature $T_{\rm gc} = 1.209\epsilon$ (Fig. 6a). Such a large jump is also observed in the force-induced transition (Fig. 6b). A non-zero bending stiffness therefore is able to dramatically enhance the cooperativity of the globule–coil phase transition. This may be partially understood in the following way. A positive bending energy significantly decreases the configurational entropy of a coil segment. Consequently the globule–coil transition will occur at higher temperature and higher force, and once the polymer is unfolded it favors those highly elongated configurations which have fewer bends.

It is interesting to notice that a non-zero bending stiffness of the polymer changes the nature of the collapse transition from being second-order to being first-order. This conclusion is in agreement with an earlier exact enumeration study [27], and it is also consistent with the mean-field calculations of Orland and co-workers [31]. It is well known that, in the 2D Ising model the paramagnet–ferromagnet phase transition changes from being second-order to being first-order under the action of a non-zero external magnetic field. What surprises us is that, in the polymer system such a qualitative change is caused not by an external field (such as the external force), but by an internal (microscopic) parameter, the bending stiffness Δ. At this point, additional MC simulation work is needed for the study of more realistic models, e.g., a self-attractive and semiflexible self-avoiding 2D or 3D chain. It is also highly desirable to perform real 2D polymer collapse experiments. For

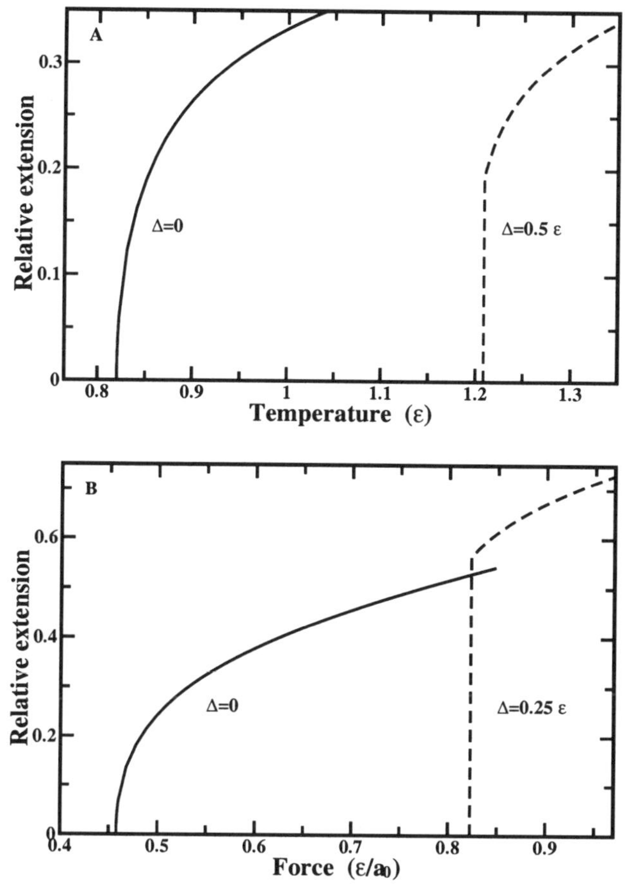

Figure 6. β-sheet unfolding. (A) temperature–extension curve at force $f = 0$; (B) Force–extension curve at temperature $T = 0.590928\epsilon$. Solid lines correspond to a flexible polymer ($\Delta = 0$), dashed lines correspond to a semiflexible polymer ($\Delta > 0$).

example, a long polypeptide chain, say poly(Glycine), can be attached to a mobile lipid bilayer [32] and its configurations can be recorded in real-time and manipulated by controlling temperature or external force.

6. Elasticity of spider capture silk

The capture silk is a natural material produced by orb-web weaving spiders. Spiders rely on it to entrap flying preys [33]. Like the spider dragline silk and many other naturally occurring silks, the capture silk has a high tensile strength that is comparable to steel; but unlike steel, it is also extremely elastic, with the ability to be stretched to almost ten times its relaxed contour length without breaking [34,35]. This perfect combination of strength and extensibility conveys a high degree of toughness to the capture silk: its breakage energy per unit weight is more than twenty times that of a high-tensile steel [34]. On the other hand, the mechanism behind spider silk's remarkable strength and elasticity is still largely missing, partly because of the difficulty to obtain high-quality crystallized structures of silk proteins.

In a recent experiment, Hansma and co-workers [35] attached capture silk mesostructures (probably composed of a single protein molecule) or intact capture silk fibers to an atomic force microscopy tip and recorded the response of the samples to external stretching force. They found a remarkable exponential relationship between the extension x and the external force f,

$$f \propto \exp(x/\ell), \tag{15}$$

where the length constant ℓ is a fitting parameter. In the spider capture silk experiment, the exponential behavior was observed at both fluid and air within a force range from about 10^0 piconewton (pN) to about 10^6 pN [35]. The exponential force-extension curve is significantly different from the predictions of simple polymer models as reviewed in Sec. 2.

Equation (15) indicates the following: (*i*) Because the capture silk is highly extensible, a great amount of extra length must have been stored in its relaxed form. (*ii*) Since extension increases with force logarithmically, some fraction of the stored length must be easy to be pulled out, some fraction be harder to be pulled out, and till some other fraction be even harder to be pulled out. To model this kind of heuristic cascading responses, a *hierarchical chain model* was suggested for spider capture silk in Ref. 36 (see Fig. 7). In the hierarchical chain model, the polymer is composed of many basic structural motifs; these motifs are then organized into a hierarchy, forming structural modules on more and more longer length scales. At the deepest hierarchy level h_m, the structural motifs could be β-sheets, β-spirals, helices or microcrystal structures. The interactions among some of these motifs are much more stronger than their interactions with other motifs, therefore they form a structural module at the hierarchy level $(h_m - 1)$. These level-$(h_m - 1)$ modules are then merged into level-$(h_m - 2)$ modules through their mutual interactions. This merging process is continued; and finally at the global scale, the whole spider silk string is regarded as a single module of the hierarchy level $h = 0$.

When the spider capture silk is under an external stretching, the total extension of a structural module at the hierarchy level $h = 0$ can be decomposed into two parts. First, the weak bonds between those level-$(h = 1)$ sub-modules of this module may break; the relative positions of these sub-modules will then be displaced, leading to an elongation of the level-0 module. This contribution of extension will eventually saturate when all the weak bonds between these level-1 sub-modules are completely destroyed. However, there is another source of elongation, namely that each of these level-1 sub-module will have an internal deformation under the stretching. The deformation of a level-1 sub-module, in turn, can be further decomposed into two parts; .... A semi-quantitative calculation was performed in Ref. 36 based on this picture of elongation cascade. The results are shown in Fig. 7. This figure suggests that the exponentially observed exponential force-extension behavior of

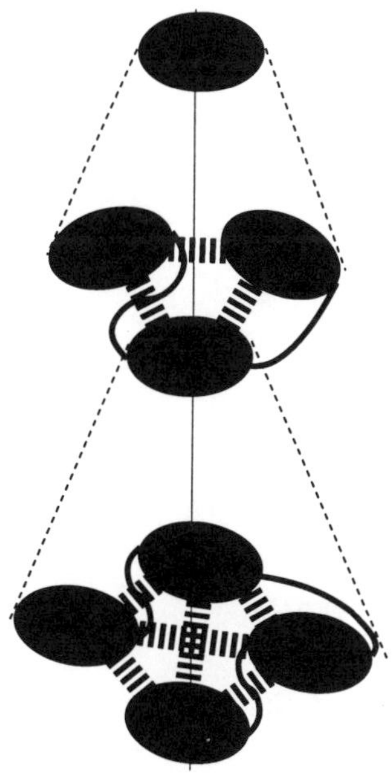

Figure 7. The hierarchical chain model for spider capture silk. At each hierarchy level h a structural module is composed of a tandem sequence of m_h submodules M_{h+1} of hierarchy level $h+1$. The thick broken lines between submodules of each hierarchy level indicate the existence of sacrificial bonds.

spider capture silk can be explained by the hierarchical chain model. According to this model, the response of the spider capture silk to external perturbations is in a hierarchical manner. If the external force is small, only those structural units of length scale comparable to the whole polymer length will be displaced and rearranged; structural units at short and moderate length scales will remain intact. As the external perturbation is increased, additional structural units at more and more shorter length scales are also deformed. Through such a hierarchical organization, a single polymer chain can respond to a great variety of external conditions; at the same time, it is able to keep its degree of structural integrity as high as possible. This hierarchical modular structure also indicates a broad spectrum of relaxation times. The modules at the shorter length scales will have much shorter relaxation times and will be refolded first when the external force decreases. This gap in relaxation times ensures that, after extension, the spider capture silk will return to its relaxed state gradually and slowly. This is a desirable feature for spider capture silk, because a too rapid contract following the insect's impact would propel the victim away from the web.

The simple hierarchical chain model, while appealing, needs further experimental validation. This model seems to be supported by recent genetic sequencing efforts. By analyzing the cDNA sequence of the major protein of spider capture silk, the flagelliform protein, it was revealed that the amino-acid sequence of flagelliform has

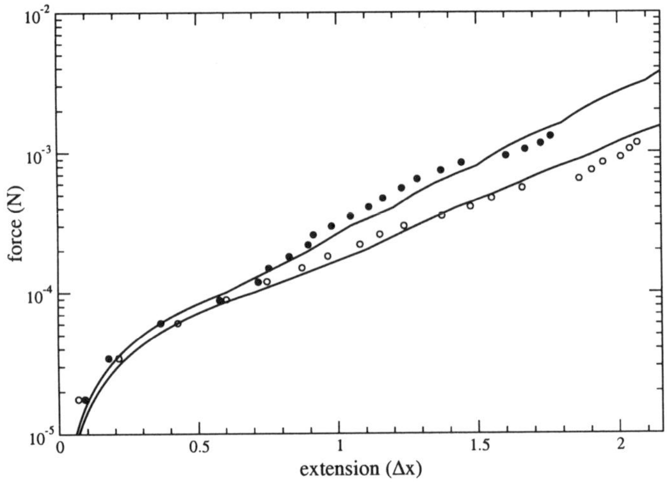

Figure 8. Exponential force-extension relationship for the hierarchical chain model. Lines are theoretical calculations [36] with two different sets of parameters and symbols are experimental data of Ref. 35.

a hierarchy of modularity [37,38,39]. At the sequence level, the structures of spider capture silks therefore have the potential to be hierarchically organized. More experimental as well as large-scale numerical simulation work are needed to fully understand the structural organization of spider capture silks.

7. Conclusion

In this review paper, we briefly described some recent theoretical work on the mechanical properties and structural transitions of biopolymers. We have discussed DNA over-stretching transition, RNA secondary-structure denature, protein β-sheet unfolding, and also structural organization principle of spider capture silk. From these studies, we get the impression that weak non-covalent bonds and interactions, convey both stability and flexibility to a biopolymer system.

Acknowledgments

I am grateful to Sanjay Kumar, Reinhard Lipowsky, Zhong-Can Ou-Yang, Yang Zhang and Jie Zhou for support and collaboration. This review paper is based on an talk given by the author at the Second International Symposium on the Frontier of Applied Mathematics, in honor of Prof. C. C. Lin at his 90th birthday.

References

1. S. B. Smith, L. Finzi and C. Bustamante, *Science* **258**, 1122 (1992).
2. H. Zhou, Y. Zhang and Z.-C. Ou-Yang, Theoretical and computational treatments of dna and rna molecules, in *Handbook of Theoretical and Computational Nanotechnology*, eds. M. Rieth and W. Schommers (American Scientific Publishers, California, 2005) pp. 1–69.
3. C. Bustamante, S. B. Smith, J. Liphardt and D. Smith, *Curr. Opin. Struct. Biol.* **10**, 279 (2000).
4. T. R. Strick, G. Charvin, N. H. Dekker, J.-F. Allemand, D. Bensimon and V. Croquette, *C. R. Physique* **3**, 595 (2002).
5. M. C. Williams and I. Rouzina, *Curr. Opin. Struct. Biol.* **12**, 330 (2002).
6. C. Bustamante, Z. Bryant and S. B. Smith, *Nature* **421**, 423 (2003).
7. J.-F. Allemand, D. Bensimon and V. Croquette, *Curr. Opin. Struct. Biol.* **13**, 266 (2003).
8. F. Bueche, *Physical Properties of Polymers* (Interscience, New York, 1962).
9. P. J. Flory, *Statistical Mechanics of Chain Molecules* (Interscience, New York, 1969).
10. J. F. Marko and E. D. Siggia, *Macromolecules* **28**, 8759 (1995).
11. C. Bustamante, J. F. Marko, E. D. Siggia and S. Smith, *Science* **265**, 1599 (1994).
12. S. B. Smith, Y. Cui and C. Bustamante, *Science* **271**, 795 (1996).
13. P. Cluzel, A. Lebrun, C. Heller, R. Lavery, J.-L. Viovy, D. Chatenay and F. Caron, *Science* **271**, 792 (1996).
14. H. Zhou, Y. Zhang and Z.-C. Ou-Yang, *Phys. Rev. Lett.* **82**, 4560 (1999).
15. H. Zhou, Y. Zhang and Z.-C. Ou-Yang, *Phys. Rev. E* **62**, 1045 (2000).
16. H. Zhou and Z.-C. Ou-Yang, *Modern Phys. Lett. B* **13**, 999 (1999).
17. M. Rief, H. Clausen-Schaumann and H. E. Gaub, *Nature Struct. Biol.* **6**, 346 (1999).
18. B. Maier, D. Bensimon and V. Croquette, *Proc. Natl. Acad. Sci. USA* **97**, 12002 (2000).
19. A. Montanari and M. Mézard, *Phys. Rev. Lett.* **86**, 2178 (2001).
20. H. Zhou, Y. Zhang and Z.-C. Ou-Yang, *Phys. Rev. Lett.* **86**, 356 (2001).
21. H. Zhou and Y. Zhang, *J. Chem. Phys.* **114**, 8694 (2001).
22. Y. Zhang, H. Zhou and Z.-C. Ou-Yang, *Biophys. J.* **81**, 1133 (2001).
23. D. Poland and H. A. Scheraga, *Theory of Helix-Coil Transitions in Biopolymers: Statistical Mechanical Theory of Order-Disorder Transitions in Biological Macromolecules* (Academic Press, New York, 1970).
24. R. Brak, A. J. Guttmann and S. G. Whittington, *J. Phys. A: Math. Gen.* **25**, 2437 (1992).
25. A. L. Owczarek and T. Prellberg, *Physica A* **205**, 203 (1994).
26. A. Rosa, D. Marenduzzo, A. Maritan and F. Seno, *Phys. Rev. E* **67**, p. 041802 (2003).
27. S. Kumar and D. Giri, *Phys. Rev. E* **72**, p. 052901 (2005).
28. S. Lifson, *J. Chem. Phys.* **40**, 3705 (1964).
29. H. Zhou, J. Zhou, Z.-C. Ou-Yang and S. Kumar, Collapse transition of two-dimensional flexible and semiflexible polymers, Unpublished, (2006).
30. P. Grassberger and H.-P. Hsu, *Phys. Rev. E* **65**, p. 031807 (2002).
31. S. Doniach, T. Garel and H. Orland, *J. Chem. Phys.* **105**, 1601 (1996).
32. B. Maier and J. O. Rädler, *Phys. Rev. Lett.* **82**, 1911 (1999).
33. F. Vollrath, *Sci. Am.* **266**, 70 (1992).
34. J. M. Gosline, P. A. Guerette, C. S. Ortlepp and K. N. Savage, *J. Exp. Biol.* **202**, 3295 (1999).
35. N. Becker, E. Oroudjev, S. Mutz, J. P. Cleveland, P. K. Hansma, C. Y. Hayashi, D. E. Makarov and H. G. Hansma, *Nature Materials* **2**, 278 (2003).

36. H. Zhou and Y. Zhang, *Phys. Rev. Lett.* **94**, p. 028104 (2005).
37. P. A. Guertte, D. G. Ginzinger, B. H. F. Weber and J. M. Gosline, *Science* **272**, 112 (1996).
38. C. Y. Hayashi and R. V. Lewis, *J. Mol. Biol.* **275**, 773 (1998).
39. C. Y. Hayashi and R. V. Lewis, *Science* **287**, 1477 (2000).

THE STRUCTURE, EVOLUTION AND INSTABILITY OF A SELF-GRAVITATING GASEOUS DISK UNDER THE INFLUENCE OF PERIODIC FORCINGS

CHI YUAN

Institute of Astronomy & Astrophysics
Academia Sinica
P.O. Box 23-141, Taipei, Taiwan
E-mail: yuan@asiaa.sinica.edu.tw

Spiral structure is the most distinguished feature common in all astrophysical disks. One of the ways the spirals can be generated is through an external periodic force in a mechanism known as the resonance excitation. We will use numerical simulations to demonstrate this process for galactic disks. A rotating bar potential or a potential due to spiral waves, both of stellar origin, acts as a periodic forcing to be imposed on a gaseous disk in a disk galaxy. We will show how the spiral density waves are generated, the mass of the disk is re-distributed, and the instability results. In other words, we show in simulations the structure, evolution and instability of the disk subject to such a periodic forcing. The instability which leads to turbulence and chaos of the disk can be identified to be of the types of Rayleigh's shear instability and Toomre's gravitational instability, or a combination of both. The results are generally shown in movies and some analyses are given. The work is in parts supported by a grant from National Science Council, Taiwan, NSC94-2752-M-001-002-PAE.

1. Introduction

Disks are the second most common configuration in the Universe. They range from 10^{10} cm for planetary rings, to about 10^{15} cm for proto-planetary disks and 10^{13-16} cm for various kinds of accretion disks, and to 10^{22} cm for galactic disks. Their sizes span over 10^{12} in magnitude. Yet they all share the same spiral structure in their appearance. A little more than forty years ago, C.C. Lin first demonstrated that these spirals are waves in galactic disks and opened a new era of galactic study and later this new era was extended to the study of all astrophysical disks (Lin and Shu 1964; 1966). Exact forty years ago, I had a great luck to become his postdoc to work on this new theory of spiral density waves. Astrophysical world has opened to me and fascinated me ever since, and spiral density waves in astrophysical disks are still my most favorite subject.

There are two ways to make spirals in a disk system. One is through a self-excitation mechanism, in which natural modes of spiral form of the disk system would appear. (e.g., Thurston et al 1989; Bertin et al 1989). Another way that the spirals can be created is through a periodic forcing in a mechanism known as the resonance excitation. Pertaining to the disk systems, such periodic forcings are naturally in existence. They are moons for planetary rings, planets for protoplanetary disks, asymmetrical central objects or mass for accretion disks, and bars for galactic disks.

The theory of resonance excitation was developed for understanding the structure of Saturn's rings (Goldreich & Tremaine 1978) and galactic structure (Goldreich & Tremaine 1979). These are linear asymptotic theories. They are great in revealing the underlying physical mechanisms which produce the large-scale spiral structure. But they fail to answer questions about the evolution of disks, and non-linear behaviors which lead to shock waves, chaotic sub-structure and violate instabilities. All of them are closely related to the fascinating observations of star formation, starburst, and energetic activities in the galactic centers and elsewhere. Subsequent non-linear asymptotic theory manages to explain the shock formation and other non-linear structural features (Shu et al 1985; Yuan & Cheng 1991). They too fail to answer the rest of the questions. It thus becomes necessary that we must resort to numerical simulations.

For numerical simulations, there are n-body methods and gas-dynamics methods. We are using gas-dynamics since it is most relevant to the disk problems. For the gas-dynamic approach, an enormous amount of literature already exists. For the problem of disk galaxies driven by a rotating bar, the earliest work can be traced to the late 70's (Huntley 1978; Roberts et al 1979). Some important results can be found in the summary of an impressive work by Athanasoula (1992). More recently, bar-driven density waves results are reported Regan et al (2005). Almost all these works, they study only a slowly rotating bar, thus exciting waves at an inner Lindblad resonance (ILR). Furthermore, they use the polar coordinates and thus need an inner boundary near the center, which introduces numerical problems. And they also do not consider the self-gravitation of the disk, therefore cannot properly address the stability problem of the disk.

In this paper, we will use the Antares codes, which we have developed for the last four years. They are high-order Godunov codes, based on the idea to calculate the flux on the interfaces, using the exact Riemann solution. They are written both in Cartesian and polar coordinates. The gas-dynamic codes are coupled with fft Poisson solver to include the self-gravitation calculation of the disk. We will present simulations of spiral density waves in a gas disk either excited by a bar potential, or responded to an imposed spiral potential, both of stellar origin and rotating as a rigid body. We can see how the spiral waves are formed, become non-linear, develop into shocks, and eventually result in instability and chaos. Both Toomre's instability and shear instability can be identified.

The presentation is organized as follows: In section 2, we briefly introduce the model rotation curve and the bar potential used in this report. In section 3, we show the results due to a fast rotating bar and their relevance to the recent high-resolution observations of galactic central regions, such as starburst rings, the circumnuclear molecular disks (CNMD). In section 4, we present the results of slowly rotating bar potential. They are relevant to the open spiral structure of galaxies such as NGC5248 and the straight-lane phenomenon in the major barred galaxies, such as NGC1097 and NGC1300. In section 5, A two-arm spiral potential of stellar origin is imposed on the gaseous disk of a galaxy. Simulations shows the development of the global pattern of doubly periodic shocks, high harmonic components of the waves and the evolution of the gaseous disk. Instability occurs in all simulations, if the imposed potential is sufficiently strong. Toomre's instability and shear instability of Rayleigh's type are observed. General remarks on some of the physical and numerical issues are made in the conclusion, section 6.

2. Rotation Curve and Bar Potential

For the simulations discussed in this report, we adopt a nearly flat rotation curve, which is

$$V = V_0\left(\frac{r}{r+\epsilon}\right)^{1/2}, \tag{1}$$

with $\epsilon = 0.01$. It rises rapidly from the center, like that of the Milky Way, representing high concentration of mass in the center. In this case, the $\Omega - \kappa/2$ curve does not have a local maximum. The horizontal line representing the pattern speed of the bar, Ω_p, would intersect with $\Omega \pm \kappa/2$ curves, resulting two Lindblad resonances, the ILR and OLR, as shown in Figure 3.

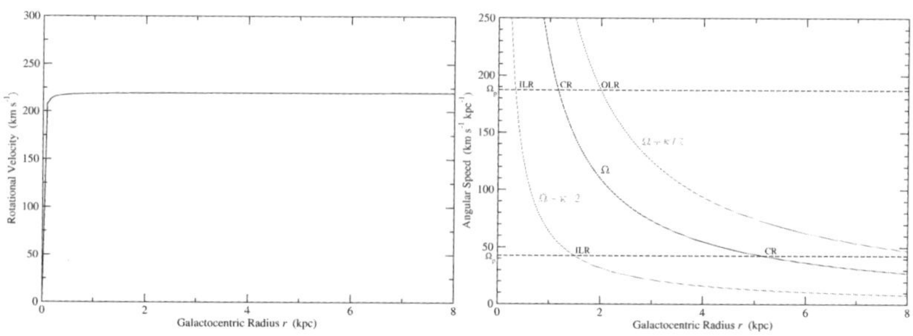

Figure 1. Nearly flat rotation curve (fast rising). The left panel is the rotation speed in km/s vs. radius in kpc. In the right panel, the middle curve is angular velocity Ω, and the top curve and bottom curve are respectively the $\Omega + \kappa/2$ and $\Omega - \kappa/2$, all in km/s/kpc. The two horizontal lines represent speeds of bar rotation. The intersection of ILR and OLR by the upper line are to be considered here.

The bar potential is taken to be

$$V = \Psi \cos(2\theta), \tag{2}$$

with

$$\Psi = \Psi_0 \frac{r^2}{(a^2 + r^2)^2}, \tag{3}$$

where a is at the potential minimum. This potential has the property that it goes to zero as r^2 and approaches r^{-2} for large r. Thus, the bar force at $r = 0$ is zero, while it behaves as r^{-3} (not r^{-2}) when r is large. In other words, the axisymmetric component of the bar potential is not present.

The initial gas density is set to be constant for the cases of non-self-gravitating disks, in which we simply take $\sigma/\sigma_0 = 1$, where σ is the surface density. For the self-gravitating disks, we use either constant or,

$$\sigma = \sigma_0 e^{-(\frac{r}{r_0})^2}, \tag{4}$$

where σ_0 is the initial surface density at the center and usually we take a value of 50 $M_\odot - pc^{-2}$. The value of r_0 is specified by $\sigma = (1/5)\sigma_0$ at $r = 3$ kpc.

3. Spiral Waves excited by a Fast Rotating Bar

A fast rotating bar, there will excite two sets of waves, one at the OLR, which is located around 1.5-3.0 kpc, and the other at ILR close to the center. The result is that a pair of tightly wound spirals at the OLR and a pair of open spirals at the ILR are excited, exactly as the asymptotic theory predicts. Since at OLR, the bar transports angular momentum to the disk. The disk material near the resonance gains angular momentum and moves out. It eventually clear a gap behind it. At the same time, under the influence of the spirals at the ILR, the disk material there will lose angular momentum and move in. After a few turns of the bar, the spiral waves at the OLR will steepen and form spiral-ring-like feature in close resemblance to the starburst ring seen in NGC4313. Near the center, an oval disk with open spirals imbedded in it is formed. Between them is a wide region void of gas.

However, when self-gravity of the disk is included, the spirals outside become unstable and develop into chaos. This is because the rapid increase of surface density, σ, in the narrow spiral-ring region forces the Toomre's $Q = a\kappa/\pi G\sigma$ there to go under 1 and thus turn that region into instability. The instability leads to starburst activities in the spiral-ring structure. This is the case for NGC1068 (Bruhweiler et al 2001) and for the Milky Way with the 3-kpc arm outside (Yuan & Cheng 1991) and a dense CNMD at the center (Jackson et al 1996). On the other hand, the central oval disk at the center is gravitationally stable even with extremely high surface density, upto 10^3 $M_\odot/pc^2$. This is because the high values of epicyclic frequency κ near the center cancel out the high surface density so Toomre's Q, defined as $Q = \kappa a/(\pi G\sigma_0)$, remains greater than 1. So it survives in the self-gravitation case. The result gives a reasonable explanation of the origin of the CNMD observed in many nearby galaxies.

The above scenario is for the case that the OLR and ILR are well separated. No interaction exists between the waves excited at the two locations. Nevertheless, when the OLR and ILR are not so far away, they may weakly interfere and give rise to the diamond-shape feature between the two resonances. This will be relevant to the observation of the double rings in NGC6782. Given the limited space here, the result is not shown here.

Shown in Figure 2 are the results of simulations for the combination of both OLR and ILR, with and without disk self-gravitation. We use a rotation curve slightly different from the nearly flat rotation in figure 1, in order to have OLR situated at 3 kpc and ILR at 0.5 kpc, more or less to simulate the Milky Way. All three phenomena are present: Starburst ring, the dense CNMD and a depression ring of gas between them.

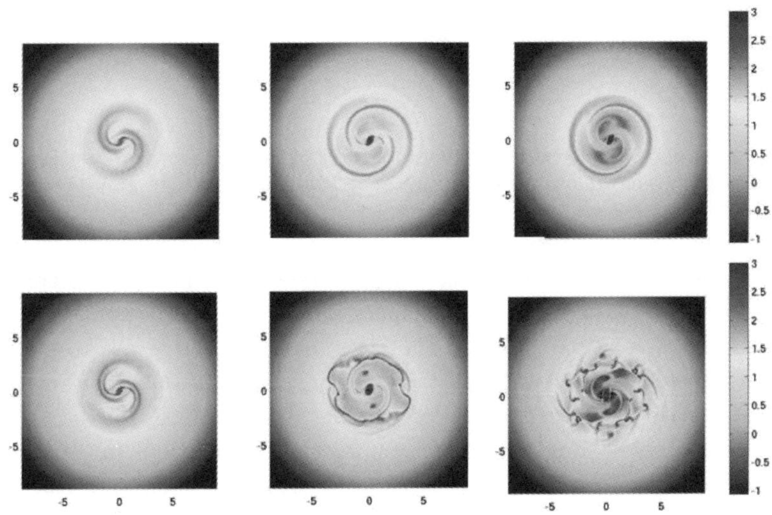

Figure 2. Case for the OLR-ILR combination. Spiral patterns excited by a bar at OLR (at 3 kpc) and at OILR (at 0.5 kpc) are shown here at 1, 4 and 8 turns of the bar. The top panels are for non-self-gravitation while the bottom panels for self-gravitation. the inner oval disk can be identified as the observed dense CNMD, which is stable even in the self-gravitational case. The outer spiral-ring structure develops into starburst rings. Also notice the gas depression gap between OLR and ILR.

The instability at OLR can be identified to be of Toomre's kind. In figure 3, we plot the Toomre's Q value at the onset of instability, which occurs at the 3rd turn of the bar. We can see it coincides with Q when it dips under 1 at $r = 3\ kpc$. On other hand, in the center where the CNMD is located, the surface density can go as high as $10^3\ M_\odot/pc^2$. The disk remains mainly stable, except along the spirals where shock waves occur.

4. Spiral Waves Excited by a Slowly Rotating Bar

If the bar potential rotates with a low pattern speed, the ILR is now located further out from the center and the OLR would be displaced to a distance no more in the galactic disk (See the lower horizontal line in Figure 1). Thus, we are facing a single Lindblad resonance (ILR) problem. The results are sensitively dependent on the

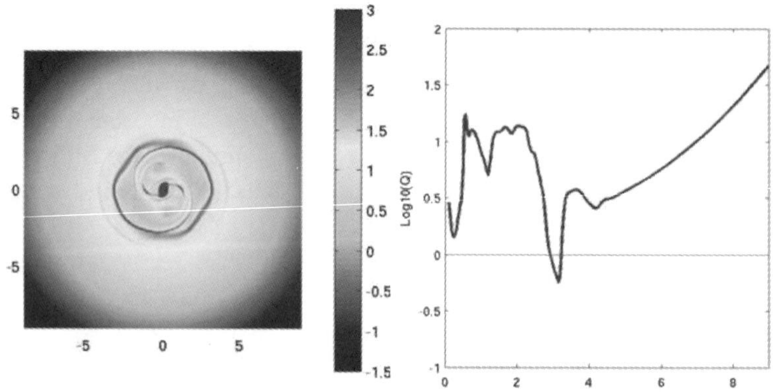

Figure 3. Onset of instability at time equal to 3 turns of the bar. Left is the surface density and right is the Q-value average on a circular ring. The instability starts to grow once Q dips below 1, or 0 on logarithmic scale

strength of the bar field. A weak bar field will result in a pair of two gentle spirals, extended from the outer parts of the galaxy all the way to the center, like the case in NGC5248. A strong bar field, on the other hand, will give rise to a pair of long straight shocks with the outer end attached to a spiral and the inner end to a bright ring. It is the case in the major barred galaxies, such as NGC1300 and NGC1097.

In figure 4, we will show the result of numerical simulations in comparison with the observations and the non-linear asymptotic results (Yuan and Yang 2006). In figure 5, the comparison of a strong bar simulation with observation of NGC1300 is shown. What we did not show is the instability which occurs in both non-self-gravitation case and self-gravitation case. For the self-gravitation case, the instability is of Toomre's type. For the case without self-gravitation, the instability is of the Rayleigh's shear type. The latter can be better illustrated in the problem of the formation of doubly-periodic shocks, which we discuss in the following section.

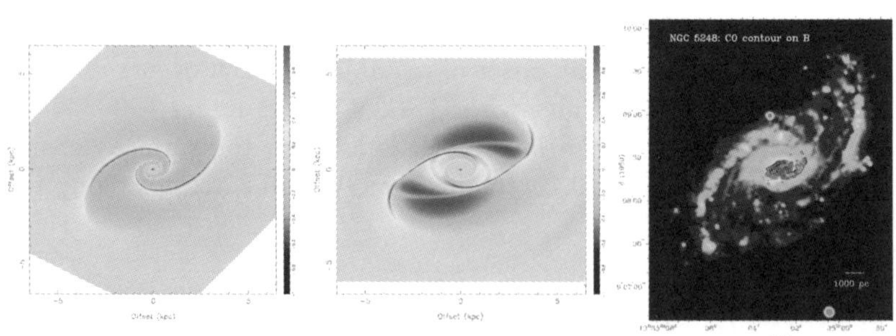

Figure 4. NGC5248: Comparison with observations. On the left is the spiral structure obtained by asymptotic analysis, center, by numerical simulations and to the right, the optical observations

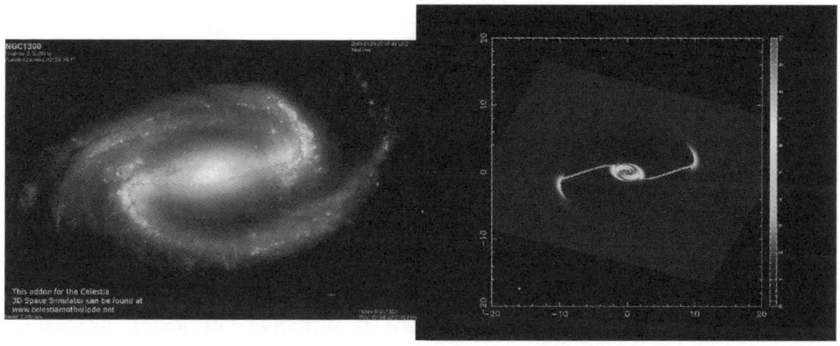

Figure 5. NGC1300: Comparison with observations. On the left is the HST high resolution observations and to the right, the simulation results

5. Doubly-Periodic Shock Solutions

5.1. *Background*

One of the remarkable results of the spiral density wave theory is the demonstration of the existence of a doubly-periodic shock solution. It satisfactorily resolves the outstanding problem of star formation along the spiral arms in disk galaxies. The seminal work of Roberts (1969) on this problem and the subsequent work (Roberts and Yuan 1970; Shu et al 1972), however, are all based on an asymptotic analysis, which is steady-state, one-dimensional and without the inclusion of self-gravity of the gas. Thus, this approach, standing high as it is, is unable to study the evolution and stability of the disk. One-dimensional time-dependent numerical computation for the problem formulated by Roberts was carried out by Woodward (1975). Although superficially it seems just to provide a time-dependent solution and confirm Roberts' results and serves no other purposes, it points out a new direction for theoretical astrophysics, or at least for theoretical galactic study, i.e., to use numerical methods to solve the full nonlinear gas-dynamic equations. Numerical simulations for two-dimensional galactic disks, using gas-dynamic codes, started late 70's. They are mainly for the bar driven problems. The two-dimensional problem of doubly-periodic shocks of Roberts' was picked up again only recently by Chakrabarti et al (2003).

Beside the main problem of star formation along the spiral arms, there are other issues. One of them is the high harmonics, which were noted in Roberts' pioneer work (1969) and late studied in more detail by Shu et al (1973). The occurrence of high harmonics was suspected to be the origin of the observed multiple arms in the outer parts of some disk galaxies. They are also believed to the sub-structure developed into instability and chaos which are commonly observed with nowadays high resolution and multi-wavelength observations. The high harmonics, instability and chaos are the topics treated by Chakrabarti et al (2003). They are also to be discussed by us here.

5.2. *Numerical One-dimension Asymptotic Theory*

We follow Woodward's approach to solve Roberts' problem numerically, with state of art numerical methods available today. Furthermore, we also include the self-gravitation of the disk with an asymptotic solution of the Poisson equation. In the case without self-gravitation of the disk, we are able to obtain the second harmonic shocks and even third harmonic shocks. The results are shown in figure 6. When the effect of the self-gravitation is included, contrary to the results of Chakrabarti et al (2003), the high harmonic shocks and the the high harmonic components, are suppressed. The results are shown in figure 7. We use the same parameters used by Woodward (1975) in the calculations, in which he adopted 1965 Schmidt model with the Sun located at 10 kpc. All the calculations are carried out on the solar circle at 10 kpc. higher pattern speeds would move the sun closer to the co-rotation, hence more into the high harmonic regions.

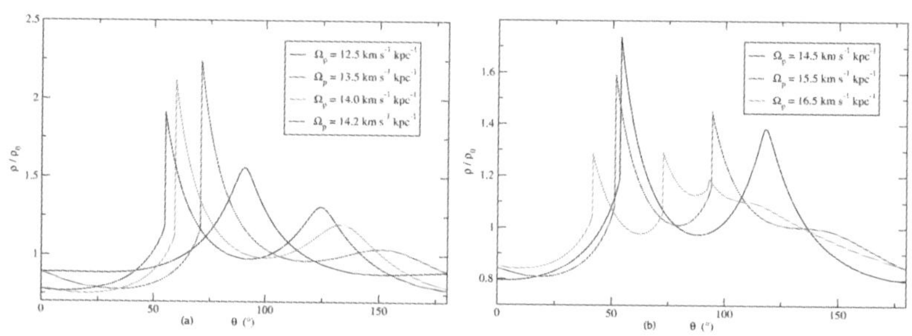

Figure 6. High harmonics. The calculations are performed at the solar circle, using the 1965 Schmidt model. Higher the pattern speed Ω_p means that the sun moves closer into the high harmonic regions.

5.3. *Two-Dimensional Numerical Results*

The problem of the gas disk in response to a two-arm spiral potential is fundamentally different from that to a bar potential. In the latter, waves are excited at Lindblad resonances and propagate as *free* waves inward or outward from the corotation. In the former, the doubly-periodic waves are not excited at the Lindblad resonances. They are *forced* oscillations driven by the imposed spiral potential.

In our calculations, we use imposed spiral fields with their field strength gradually reducing to zero before reaching the Lindblad 2:1 resonances (OLR and ILR). In doing so, it would eliminate the unwanted contributions of resonantly excited waves which may arise at the resonances. This makes our computation very different from that of Chakrabarti et al (2003), whose results are strongly contaminated by the resonantly excited waves. Furthermore, we pay special attention to the case of the doubly-periodic spiral waves within the corotation circle, which was indeed the case

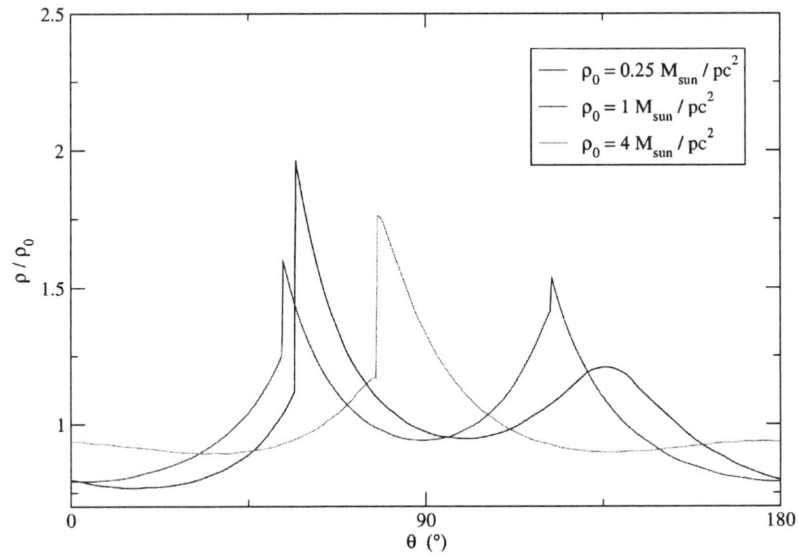

Figure 7. High harmonic components suppressed when the self-gravitation of the disk is included. Different surface densities of the gas are used. The secondary shock and hump disappear when the surface density is equal to 4 $M_{\odot} ps^{-2}$

studied by Roberts and Woodward, in addition to the general case in which spiral waves are considered between the OLR and ILR, therefore across the corotation. The results can be recognized in terms of those of Roberts or Woodward, but the global pattern, as it evolves, is far more complex than theirs.

We show first the evolution of the disk for the case of a moderate forcing within the corotation circle in figure 8. In the second case, we show the structure of the waves which covers the entire disk from the OLR to the ILR. For the latter, the imposed spiral field varies like a Gaussian distribution, centered at the corotation and with standard deviation of 5 kpc. A snapshot of the disk after the imposed spiral pattern has made 6 turns, is shown in figure 9. In both cases, we adopt a nearly flat rotation curve, the corotation is situated at 12 kpc, the imposed spirals are of logarithmic type with pitch angle equal to 18°, and the self-gravitation of the disk is not considered.

5.4. *Instability and Chaotic Sub-structure*

When the strength of the spiral field is increased, the doubly-periodic shocks becomes unstable. The onset of instability seems to occur when the shock strength is sufficiently high, independent what has caused it. Therefore the same results are achieved by merely reducing the grid size, or increasing the resolution. In figure 10, we show how the instability develops. A number of pockets of high-low vorticity pairs form along the shock, and they gradually grow in size, become sub-structural patches and eventually move downstream from the shock along the instantaneous

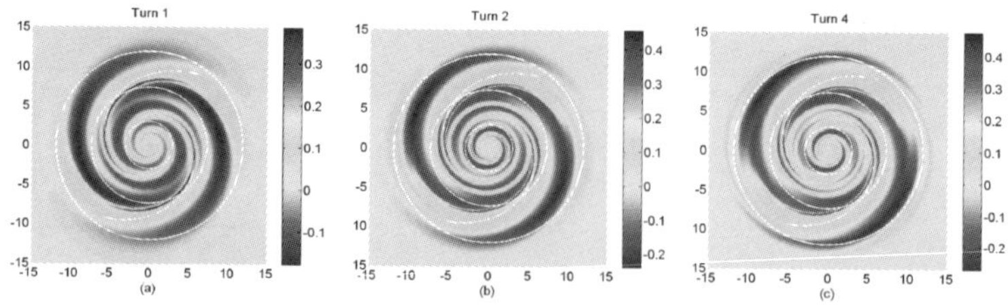

Figure 8. Density evolution of the disk under the imposed two-arm spiral potential, shown in white color. The three white circles are from outside corotation, 4:1 ILR, and 2:1 ILR. The doubly periodic shocks lie inside of the imposed spirals (color white). The 2nd harmonic components are excited at the 4:1 ILR, propagating inward as theory predicts. There is a residual two trailing spiral excited at 2:1 ILR propagating toward the center.

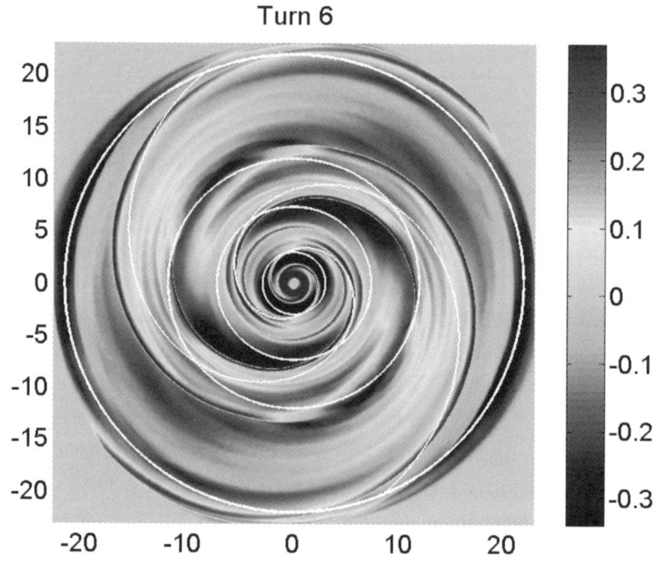

Figure 9. Snapshot of density distribution of the disk after 6 turns of the imposed spiral potential for the entire disk from OLR to ILR. Again, shown in white color are the imposed spirals, four circles respectively the OLR, corotation, 4:1 ILR, and 2:1 ILR. High harmonic components are clearly seen especially inside the corotation.

streamlines. They will eventually populate along a bundle of streamlines to form a ring of chaotic patches.

We believe the instability belongs to the type of Rayleigh's shear instability. Lord Rayleigh has shown that the criterion for instability of a parallel flow of inviscid incompressible fluids is when the velocity profile of the flow has a point of inflection, or the vorticity has a local extremum. Translating that criterion for an axisymmetric rotating flow, it becomes that the specific vorticity, has a local extremum. Here the specific vorticity is defined as vorticity divided by density. This is because the specific vorticity plays the same role in rotating flows of compressible fluids as vorticity in the parallel flows of incompressible fluids. For two-dimensional flows, the vorticity for the latter satisfies the vorticity conservation equation,

$$\frac{D\omega}{Dt} = 0,$$

while the specific vorticity follows,

$$\frac{D\omega}{Dt}(\frac{\omega}{\rho}) = 0,$$

where ω is the vorticity and rho, the density. The physical mechanism was first explained by C.C. Lin (1944). The above equations, however, are valid if there is no shock. If shocks occur, vorticity and hence specific vorticity are created (See e.g., Shu 1992). In figure 11, we calculate the specific vorticity by using central difference of the velocity field obtained numerically. It clearly demonstrates that the wiggles along the spiral shocks in the density plot are local specific vorticity extremum pairs. By Rayleigh's criterion, they are the pockets which have the potential to develop into instability. This indeed takes place in the calculations. Furthermore, in compressible fluids, unlike the sound waves, entropy-vortex disturbances would follow the fluid motion. This explains why we see the sub-structural patches moves along the streamlines, eventually forming a ring-like structure. In figure 10, we show the onset of instability along a spiral shock. The left panel is the density distribution, the middle, the specific vorticity calculated by central difference of the velocity field in the simulation, the left, the specific vorticity calculated as if there were no shock.

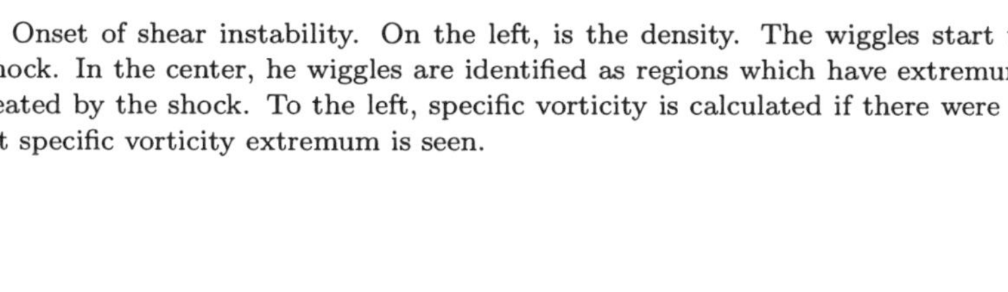

Figure 10. Onset of shear instability. On the left, is the density. The wiggles start to appear along the shock. In the center, he wiggles are identified as regions which have extremum specific vorticity created by the shock. To the left, specific vorticity is calculated if there were no shock. No apparent specific vorticity extremum is seen.

5.5. *Effect of Self-gravity*

When the self-gravitation of the disk is include, Toomre's instability appears especially in the outer parts of the galaxies where the epicyclic frequency is low, therefore inertia force is small. This is shown in figure 11. A gas disk which is normally stable against the shear instability when a spiral field which has the strength of 3% of the main field at 8 kpc, develops into instability in the outskirts of the disk.

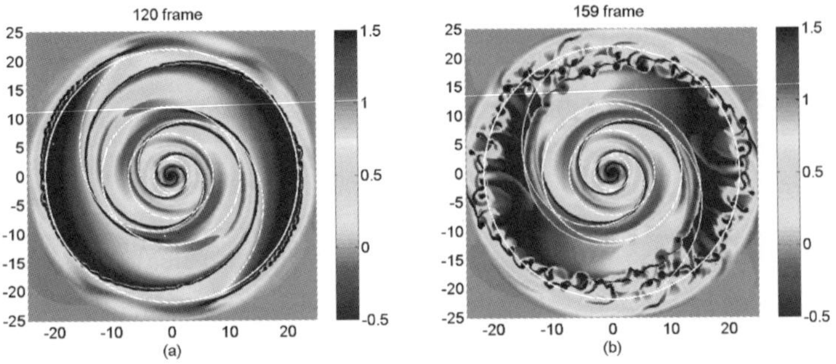

Figure 11. Toomre's instability. When the self-gravitation of the disk is include, a shear-stable disk becomes unstable. The instability occurs in the outskirts of the disk, where the inertia due to rotation is weak. The onset of stability appears after 1.5 turns of the imposed spiral density potential and chaos is fully developed after another half turn.

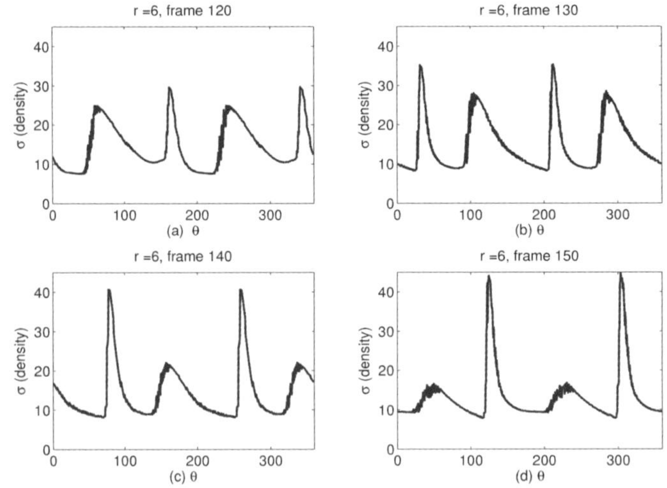

Figure 12. Effect of self-gravitation on high harmonics. The self-gravity tends to enhance all wave components. In the end, the enhance of the primary shock component overshadows the rest of the high harmonics.

The self-gravitation in general helps to organize the spiral structure and to trim the spiral arms, as long as the instability does not appear. In this sense, it tends to suppress the high harmonics as we show earlier in the one-dimensional case. However, the situation is more complicated. It actually first enhances all the harmonic components. Eventually, the enhancement of the primary component is so predominant that it takes over the rest of the high harmonics. Figure 12 demonstrates this process.

6. Concluding Remarks

We have shown that the starburst ring, the dense CNMD, and a gas-depression region between them in the central region of a disk galaxy can be produced by a fast rotating bar potential. The starburst ring can be identified as a phenomenon associated with the OLR, while the dense CNMD, the ILR. The gas-depression region, between them is a result of gap clear by the spiral density waves excited at the two resonance. Another important result of this study is the identification of Toomre's instability. It natural explains why starburst should occur and CNMD should exist.

If the bar potential rotates slowly such that the ILR is located outside the central region, say 5 kpc, a pair of open spiral arms would be excited, which can cover the entire disk from the outer most parts extended all the way to the center. This is the case of NGC5248. If the bar strength is strong, a pair of straight line shocks will form. Towards the center, they connects to a circular ring-like structure, which can be identified as starburst ring in the case of NGC1097, and outward to a pair of tightly wound spirals. These results match perfectly with the observed dust lanes in major barred galaxies such as NGC1300 and NGC1097.

The motivation to study the response of the gas disk to an imposed spiral gravitation field is to explain the star formation along the spiral arms in disk galaxies. The two-dimensional numerical simulations demonstrate the complex nature of this problem and enrich our knowledge of this field. The fact that shear instability of Rayleigh's type can be identified is an important contribution of the study. Now we can see how the specific vorticity is created along the shock and grow into sub-structure patches, and eventually move out along the streamlines to form a chaotic ring structure. These substructures are seen in the high resolution and multi-wavelength optical observations. Inclusion of the self-gravitation of the disk will induce instability of Toomre's type on the outskirts of the galactic disk. In the inner parts, if gas mass is moderate, the self-gravity will tend to organize the spiral structure, to enhance the primary shock and suppress the harmonics in the end.

Despite of all these results, nevertheless, we should remember that spirals are the most common features in astrophysical disks, and they are natural models of the disk system. They can be generated by a self-exciting mechanism, or by an external disturbance. As a physical problem, the former is much harder than the latter. The simplest kind of the latter would be the case of periodic disturbances, which we choose to study in this paper. We use disk galaxies as our host disk system and a rotating bar or two-arm spiral potential of stellar origin as our periodic disturbance. The disk we consider is not the entire disk of a galaxy, but only its gaseous component. It is embedded in the stellar disk system and contribute only about 10% to the total mass. For the central gas disk, driven by a fast bar, the problem is even more simplified, since the gas component, with its low sound speed, is almost completely decoupled from the high dispersion speed stellar component which can hardly form a stellar disk there. So we can treat the problem by means of pure gas-dynamic equations. However, for the problem which involves the entire galactic disk, such as density waves driven by the slow bar or spiral potential, we are not so lucky. The stellar disk and gaseous disk are coupled. Our treatment here is just an approximation. We have good reason to believe it is a good approximation, since the gas content is small. To be noted is we have only studied the two-dimensional problem. To do the problem correctly, we must combine the treatment of stellar dynamics and extend our analysis to three-dimension. This is where our future study should go.

Acknowledgments

I wish to thank David C.C. Yen, Hsiang-Hsu Wang, and Lein-Hsuan Lin for the help and contributions without which this paper can be of the present shape. The work is in parts supported by a grant from National Science Council, Taiwan, NSC95-2752-M-001-007-PAE.

References

1. E. Athanassoula, *MNRAS*, **259**, 345 (1992).
2. S. Chakrabarti, G. Laughlin, F. H. Shu,*Astrophys. J.* **296**, 220 (2003).
3. P. Goldreich, & S. Tremaine, *Astrophys. J.*, **233**, 857 (1979).
4. J.M. Huntley, R.H. Sanders, & W.W. Roberts, *Astrophys. J.*, **221**, 521 (1978).
5. J. M. Jackson, M. H. Heyer, T. A. D. Paglione, A.D. Bolatto, *Astrophys. J.*, **456**, L91 (1996).
6. C.C. Lin, *Quart. Appl. Math.*, **3**, 117 (1945).
7. M. W. Regan, & P. J. Teuben, *Astrophys. J.*, **600**,595, (2004).
8. W. W. Jr Roberts, & C. Yuan, *Astrophys. J.*, **161**, 887 (1970).
9. W. W. Jr Roberts, *Astrophys. J.*, **158**, 123 (1969).
10. F. H. Shu, "Physics of Astrophysics: II. Gas Dynamics, University Science Books", Mill Valley, CA (1992)
11. F. H. Shu, V. Milione, & W. W. Jr Roberts, *Astrophys. J.* **183**, 819-842 (1973).
12. P. Woodward, *Astrophys. J.*, **195**, 61, (1975).
13. C. Yuan, & C.C. Yang, *Astrophys. J.*, **644**, 180 (2006).

DYNAMICS OF SPIRAL GALAXIES*

G. BERTIN

Department of Physics, University of Milano, via Celoria 16
Milano, I-20133, Italy

The dynamics of spiral galaxies is a gold-mine of challenging problems for the astrophysicist and the applied mathematician. In Astrophysics, we may ask how these island-universes formed, evolved, and reached their current structure and thus address the problems of the dynamics of the interstellar medium, of star formation, of stellar dynamics, and of the presence and role of dark matter halos. These fundamental issues have attracted the interest of the scientific community for almost a century. The models conceived and developed to sharpen and to quantify our understanding of the basic dynamical processes at work in this context require a semi empirical approach and general tools that are characteristic of Applied Mathematics. In particular, the beautiful morphology of spiral galaxies poses a number of interesting questions, most of which have found a coherent answer in the framework of the Density Wave Theory. I will give a synthetic description of the main concepts and of the main achievements of the theory, as developed in the course of three decades in the second part of last century. The theory has had a major impact on Astrophysics and has been the inspiration for a number of important theoretical and observational investigations. The relatively recent advent of near-infrared observations (especially those in the K-band, probing the underlying stellar component in galaxies) has confirmed that indeed large-scale spiral arms are associated with a smooth, sinusoidal density perturbation in the stellar disk, that grand design is very frequent and generally two-armed, and that multiple-armed spiral structure is mostly associated with the gaseous interstellar medium. I will also briefly outline other interesting topics in the dynamics of spiral galaxies, where progress may take place as a result of a fruitful exchange between Astrophysics and Applied Mathematics.

1 Introduction

It is a great pleasure and a great honor for me to present this paper to celebrate Professor Lin's ninetieth birthday. I may start by briefly going back in time and by revisiting the last forty years, decade by decade. In 1966 the Density Wave Theory had just been born [1,2]. In 1976 the first key steps were taken in the direction of a theory of self-excited global modes [3,4]. In 1987, a Symposium took place at MIT to honor the career of Professor Lin [5], at a time when the key steps in the direction of a unified theory of normal and barred spiral structure had been taken [6-8]. About ten years later, a monograph was published [9], summarizing in a coherent presentation all the results obtained in the theory.

The Density Wave Theory of spiral structure in galaxies is one of the most important achievements in Astrophysics. A semi-empirical approach, based on the working hypothesis of quasi-stationary spiral structure, has originated an impressive number of quantitative observational tests that have attracted the interest of the astronomical

* This work is partly supported by the Italian MIUR (cofin-2004)

community in the last four decades. The theory develops concepts and leads to predictions that are relevant not only to the dynamics of spiral galaxies, but also to the physics of the interstellar medium and to the processes of star formation, to the role of dark matter halos, and, in general, to the problem of the evolution of galaxies. The development of a successful and internally consistent theory has required the solution of a number of challenging conceptual problems at the frontier of Astrophysics and of Applied Mathematics.

In Sect. 2, I will start by briefly commenting on the morphology of spiral galaxies. In Sect. 3, I will outline the basic structure of the theory. Here, it would be impossible to cite all the papers that have contributed to the growth of this particular research area. In addition, since the mathematical structure of the theory has been presented before, on several other occasions, the text here will be kept to a discussion of the main ideas only. Relevant mathematical material and a large number of references can be found in the monograph cited earlier [9] and in [10]. In Sect. 4, I will conclude by mentioning other interesting problems in the dynamics of spiral galaxies and by presenting some prospects for future work.

2 Morphology of spiral galaxies

The morphology of spiral structure is often characterized by impressive regularity and demonstrates that law and order can govern systems that are intrinsically extremely complex. In the 60's such beautiful morphology was captured by the pictures shown in the Hubble Atlas of Galaxies [11]; later, after great progress in telescopes and instrumentation, the beautiful images have been collected in more advanced atlases [12-14], and can now be easily retrieved on the web.

The Hubble morphological classification [15] is a relatively simple framework still in use today. A natural question posed by such classification is why galaxy disks come in two categories, with either normal or barred spiral structure, but with significant continuity between the two classes. In general, one would like to understand the origin of such simple classification. Large-scale regular structure is generally outlined by sharp dust lanes, which in barred galaxies take on the form of a pair of offset straight lanes in the central regions dominated by the bar (see the case of NGC 1300, NGC 1097, NGC 5236). Some galaxies (such as NGC 3031, NGC 4622, NGC 5194) show no trace of a bar, not even in their innermost regions. Many barred galaxies exhibit a bar in the form of a two-blob structure (see NGC 2859, NGC 7743, NGC 1398). With the exception of NGC 4622 (in which a one-armed inner structure winds out in the opposite direction with respect to that of the outer two-armed structure), large-scale spiral structure is generally trailing with respect to the overall rotation. While certain galaxies definitely show an extremely regular grand design (see NGC 3031, NGC 1350, NGC 5364, NGC 4321), others are less regular (see NGC 2997 and NGC 309), multi-armed (NGC 5457, NGC 2403), or even flocculent (NGC 2841, NGC 5055). The proximity to a nearby galaxy is often suggestive of the presence of substantial tidal interactions (see the system of NGC 5194 and NGC 5195). Some disks are definitely lopsided (NGC 5457, NGC 1637, NGC 4254). Edge-on views of spiral galaxies show that the disk is generally thin and symmetric, with the possibility of some large-scale warps in the outer parts (usually in the gas only, see NGC 5907, NGC 4565; but see the impressive optical warp observed in ESO 510-G13).

3 Density Wave Theory of spiral structure in galaxies

Galaxy disks can generally be considered as axisymmetric systems, even in the cases in which large-scale spiral structure is observed. Of course, such a symmetry is only approximately realized in nature, but kinematical studies confirm that, in the gravitational field, deviations from axisymmetry are generally small. The disk material rotates around the center along circular orbits. Galaxy disks, primarily made of stars, can thus be seen as a system of oscillators. Individual star orbits in the plane of the disk are characterized by two frequencies, Ω and κ, corresponding to rotation and epicyclic oscillations around circular orbits. Disks are rather cold, in the sense that, statistically, the star epicycles are small.

The first dynamical paradox posed by the observed morphologies is the so called *winding dilemma*: since the disk rotation is not rigid but differential, $\Omega = \Omega(r)$, any material arm would be rapidly stretched into a tightly wound spiral structure. How can we reconcile this fact with the observations of so many galaxies with open arms? A first clue to the solution of the paradox was given by B. Lindblad, who noted that, if (as it appeared) the combination $\Omega(r) - \kappa(r)/2$ were approximately constant with radius, a two-armed quasi-stationary pattern (a *kinematic wave*) could be constructed even in the presence of differential rotation. In fact, this argument was a precursor to the Density Wave Theory.

3.1. Quasi-stationary spiral structure?

As well stated by Oort [16], the problem of spiral structure can be schematically divided in two parts: (i) How did spiral structure originate? (ii) How does it persist once it has originated? Oort emphasized the importance of focusing on the structure on the large scale and argued that soon progress in the observations would have been able to determine whether spiral arms are primarily stellar or primarily gaseous. In reality, beyond his formulation of the problem, there are several additional key issues that demand an explanation. In fact, we would like to understand why certain spirals are barred and others are not, what determines the different degrees of regularity observed in galaxy disks and, in particular, why some galaxies are flocculent, why the structure is generally trailing, why the grand design is generally two-armed, why disks often exhibit coexisting morphologies, what sets the amplitude of the observed spiral structure. In the end, we would like to explain the origin of the Hubble morphological classification.

To be sure, Oort's second question basically *assumes* that the large-scale spiral structure *must* be stationary. Even if we do not have a direct proof of quasi-stationarity, very interesting consequences follow from this assumption, used as a working hypothesis. Firstly, the problem of persistence could be solved naturally if spiral arms are not thought of as *material arms*, but as *density waves*. Secondly, if indeed the large-scale spiral structure is quasi-stationary, because of the presence of differential rotation the spiral (density wave) pattern is expected to move supersonically with respect to the interstellar medium over most of the disk (it soon became clear that the corotation circle should be placed in the outer parts of the optical disk), and would thus be able to generate shocks that should trigger coherent processes of star formation. Indeed, large-scale arms are generally delineated by HII regions and young stars, much like whitecaps tracing the crests of ocean waves. Such shock scenario was soon worked out quantitatively [17] and confirmed as one of the most stringent tests of the theory [18].

In reality, there are at least four alternative scenarios that can be adopted to interpret spiral structure in galaxies, each based on a different attitude in relation to the two issues raised by Oort. (1) A number of arguments supports the scenario in which large-scale spiral structure is generally *quasi-stationary and intrinsic* (i.e., its origin is internal to the galaxy where it is observed). This working hypothesis is at the basis of [9] and has led to a coherent answer to basically all the issues that can be raised in relation to the problem of spiral structure in galaxies. From a dynamical point of view, in order to demonstrate that this picture is viable, it has been shown that a wide class of realistic galaxy disks is subject to a small number of unstable global modes. (2) An alternative scenario is that in which the origin is intrinsic but the structure is rapidly evolving and possibly recurrent [19,20]. (3) Another possibility is that of quasi-stationary structure of external origin, in which the observed structure would correspond to the tidal excitation of otherwise damped modes, part of a discrete spectrum. (4) Finally, a fourth possible scenario is that in which spiral structure is transient and is excited from the outside by an occasional fast encounter with another galaxy, under suitable circumstances [21]. So far, the last three alternative scenarios have not been brought to completion into a coherent theory, although the last scenario, especially in connection with the description of the so-called *swing mechanism*, has received a lot of attention and popularity.

3.2. Density waves

The first quantitative formulation of density waves, in the form of a dispersion relation for linear perturbations on a thin axisymmetric disk, conceived to describe the properties of large-scale spiral arms, was given by Lin and Shu [1,2]. At that time, most of the efforts were made in the direction of describing a disk made primarily of stars. Later, in the late 70's and early 80's, it became clear that the basic picture is well captured by a much simpler dispersion relation that describes the properties of density waves in a *fluid disk model* (obviously, to describe some specific processes, such as resonant effects, the fluid model should be supplemented with the results of kinetic analyses). This is a relation, quadratic in the magnitude of the radial wavenumber k (at fixed m, the sign of k distinguishes *trailing* from *leading* waves), which is derived under a suitable WKB ordering (see [9]). The dispersion relation can be studied as a relation $\omega = \omega(k)$, in the spirit of a *local stability analysis*, to find that local stability is governed by a parameter, $Q = c\kappa/\pi G\sigma$, which is basically a measure of the disk temperature; this is the analogue of a well-known parameter found in the kinetic study of a disk of stars [22]. Relatively warm disks, with $Q \geq 1$, are locally stable with respect to axisymmetric Jeans instabilities. In turn, in the spirit of a *semi-empirical approach*, under the hypothesis of *quasi-stationary spiral structure*, the dispersion relation can be studied as a relation $k = k(\omega;r)$ to *draw spiral arms consistent with the disk dynamics*; this is done by assigning the observed number of arms m and the pattern frequency $\Omega_p = \omega/m$, taken as a free parameter to be determined from the observations. This latter approach takes advantage of the simple geometrical relation that defines the pitch angle of spiral structure i in terms of the azimuthal (m/r) and radial (k) wavenumbers: $\tan i = m/(rk)$.

One ambiguity present in the latter semi-empirical approach is related to the fact that the dispersion relation generally admits, within the empirically realized context of trailing waves, two wave branches. The large number of successful observational tests that followed immediately the formulation of the basic dispersion relation for density waves were all based on the use of the so-called *short-wave branch*.

A major concern was then raised [23], about the fact that a (short) density wave packet would be bound to disappear quickly from the disk, by group-propagating inwards toward the galaxy center; thus the self-consistent density waves used to fit the observations (e.g., in [18]) could not possibly correspond to a quasi-stationary spiral structure, as assumed to begin with. In response to this concern, it was argued that the central regions of galaxy disks should be capable of returning density wave signals to the outer parts by means of a suitable feedback process [24].

3.3. The importance of gas, self-regulation, and feedback

The cold dissipative interstellar medium plays a very important role in the excitation and maintenance of spiral structure in normal spiral galaxies, but this point was initially underestimated. In the early developments of the theory, the cold gas was thought to play mostly a passive role, via the shock scenario, although it had long been known that a small amount of cold gas can have a significant impact in destabilizing the disk with respect to density perturbations [2].

Later it was realized (see [25] and references therein) that such cold gas, being dissipative, can establish a sort of "thermostat" in the disk: the disk cannot become too cold, otherwise Jeans instabilities would set in, stir it, and heat it, while, because of dissipation, disks cannot become too warm either. In other words, we should expect the disk to be *self-regulated* with the effective Q (representing the combined effects of gas and stars) close to unity in the main body of the disk and in the outer parts, that is in the parts of the disk where *locally* the gas fraction is sufficiently high for self-regulation to take place.

In contrast, in the innermost parts of the galaxy the gas fraction is insufficient and the thermostat should break down; we thus expect the disk, in terms of the effective Q, to become hotter and hotter while we move inwards toward the galaxy center. In turn, density waves cannot exist nor propagate in the part of the disk where the effective Q is substantially larger than unity. This is the physical justification for the adoption of the Q-profiles (flat profiles, with an inner "Q-barrier") that, from the late 70's on, were used and realized to possess the desired *feedback mechanism* as a built-in process. A demonstration of how nicely self-regulation is established in self-gravitating disks has been provided recently by means of numerical experiments aimed at simulating the dynamics of protostellar disks [26].

3.4. Overreflection at corotation

Another very interesting process had been discovered in the meantime. Mark [27,28] found that if a (trailing) long-wave signal is launched outwards toward corotation, with consequent transfer of angular momentum across corotation toward the outer regions, a reinforced short-wave signal is returned back to the central regions of the galaxy. Such *overreflection* (sometimes called WASER) is an important mechanism that can take place in shear flows, when transfer of energy and angular momentum couples a region characterized by negative density of wave action (the disk inside the corotation circle) with a region characterized by positive density of wave action (the outer disk, outside the corotation circle).

3.5. Discrete spectrum of self-excited global modes

The combination of feedback from the central regions and overreflection at corotation makes it possible for the disk to act as a "resonant cavity", giving rise to a discrete spectrum of unstable global modes [3,4,29]. From the mathematical point of view, the integro-differential problem that governs the linear density perturbations $\sigma_1(r,\theta,t) = \sigma_1(r)\exp[i(\omega t - m\theta)]$ in a thin fluid disk can be neatly reduced, under a suitable ordering based on the coolness of the disk, to a relatively simple second-order, ordinary differential equation in a single perturbed variable u(r), which is directly related to the density perturbation $\sigma_1(r)$. [It is possible to show that such equation, studied under the

WKB algebraic approach, contains indeed, as an asymptotic limit, the quadratic dispersion relation mentioned earlier in Sect. 3.2.]

The physical prescription of an effective Q-profile of the type described in Sect. 3.3, to incorporate the mechanism of self-regulation, then leads to a Schrödinger-like equation for the variable u(r), characterized by two turning points: an inner *simple turning point*, at $r = r_{ce}$ is responsible for the feedback of short into outgoing long trailing waves, while a *double turning point* at the corotation circle $r = r_{co}$ corresponds to the location where overreflection takes place. Solved under the natural boundary conditions of evanescent wave at $r < r_{ce}$ and of outgoing wave at $r > r_{co}$, the relevant eigenvalues ω and eigenfunctons $\sigma_1(r)$ are determined from a Bohr-Sommerfeld quantum condition. Because of overreflection, the quantum condition requires the presence of an imaginary part which determines an imaginary part in ω, corresponding to exponential growth. The growth rate of the global mode is inversely proportional to the bounce time of a wave packet along a cycle, from r_{co} to r_{ce} and back to r_{co}.

3.6. A unified theory of normal and barred spiral modes

In the early 80's it was realized [6,7] that the quadratic dispersion relation for density waves on a fluid disk is actually the limiting case of a more general *cubic dispersion relation,* which depends on a second local stability parameter J, in addition to Q. Such a cubic relation reduces to the better known quadratic relation in the limit of vanishing J. The new parameter J is independent of the disk temperature and is proportional to the equilibrium disk density. Heavier disks (disks not embedded in massive bulge-halo spheroidal components) are characterized by higher values of the J parameter; lighter disks, with smaller values of J, are well described by the quadratic dispersion relation.

The (J,Q) parameter plane is divided by a transition line, $JQ^3 = (16\sqrt{2})/27$. Below (and to the left of) this line, the cubic admits up to three real solutions in the magnitude of k and all the relevant dynamical processes take the form of those described in the previous subsections (based on short and long trailing waves). Above (and to the right of) this line, the cubic admits only one real solution in the magnitude of k, and thus the relevant mechanisms should be based on the combination of trailing and leading waves (the only wavebranches available). Indeed, it has been shown that the mechanisms that underlie the excitation of a discrete spectrum of global spiral modes described above (in terms of short trailing and long trailing waves trapped in the "resonant cavity" between r_{ce} and r_{co}) carry through and have a completely analogous counterpart in this new regime of high J, provided one replaces the role of short trailing and long trailing waves with that of trailing and leading waves. In fact, also in this new regime, the integro-differential problem can be reduced to a second-order, ordinary differential equation for a new variable w(r), related to $\sigma_1(r)$, of the Schrödinger type, then leading to a Bohr-Sommerfeld quantum condition.

In the new regime of high J the eigenfunctions are completely different and generate *barred spiral structure.* In practice, across the transition line, depending on the characteristics of the basic state of the galaxy disk, a sort of phase transition takes place, with normal spiral structure occurring for relatively light disks and barred spiral structure being developed for relatively heavy disks. A survey of more than one thousand galaxy models, for which the exact integro-differential problem has been solved numerically [8], has confirmed the robustness of the results obtained from the asymptotic analysis and has led to the identification of mode prototypes that conform to the observed morphologies, from normal spiral structure, to barred spiral structure, and to the two-blob structure characteristic of SB0 galaxies (see Sect. 2).

In view of the observations, the mode shapes (eigenfunctions) are predicted together with their corotation radii (eigenvalues). In particular, for barred modes corotation is expected to occur just outside the tip of the bar in the main disk, as indeed observed (e.g., for the SB0 galaxy NGC 936, see [30]). Instead, normal spiral modes are expected to have their corotation radii located in the outer disk. Another feature, natural in the theory of modes, that neatly corresponds to the observations is the amplitude modulation along the arms, which is related to the interference of the elementary density waves that compose the global modes (e.g., compare the images of NGC 1300 with the structure of prototypical bar modes).

3.7. Role of Inner Lindblad Resonance and of some non-linear processes

The key mechanism that restricts the discrete spectrum of unstable global modes to a very small set is thought to be that of resonant absorption in the stellar disk at the Inner Lindblad Resonance (where the pattern frequency resonates with the epicyclic motions, $\Omega_p = \Omega - \kappa/m$). In practice, the resonant cavity described in Sect. 3.5 cannot operate for higher modes (modes with higher m or with lower pattern frequencies), because such absorption at ILR interrupts the relevant wave-cycle. Typically, only few modes (with m = 1,2) remain available. For the gas, the process is less efficient and multiple-armed spiral structure can thus be generated.

The exponential growth that characterizes global spiral modes applies to the linear stage only. Observed spiral structure is thought to correspond to a situation where the growth is saturated non-linearly by dissipation in the large-scale shocks [31,32]. Cold gas is thus consumed by the disk in the combined process of self-regulation and of amplitude saturation through large-scale shocks, raising important questions about the overall long-term evolution of the disk.

Another issue of long-term evolution is raised by the fact that spiral arms and bars are associated with torques, with a net flux of angular momentum to the outer regions [33].

3.8. A unified framework for the Hubble morphological classification

What has been learned by studying the problem of spiral structure in galaxies under the hypothesis that spiral structure is quasi-stationary and intrinsic can be put together to form a unified framework for the observed spiral morphologies, based essentially on a three-dimensional parameter space [9]. It appears then that the Hubble categories from early-type (a) to late-type (c) are mostly governed by the gas content. The subdivision between normal (SA) and barred (SB) galaxies should broadly correspond to a distinction between effectively lighter and effectively heavier disks. Finally, the distinction between grand-design and flocculent spiral galaxies should correspond to the case where the star and gas components are dynamically coupled (see [25] for a detailed description) and to the case where these components are decoupled. This framework is a useful reference paradigm, but should be improved by modeling a large number of individual morphologies in the light of viable alternative scenarios.

3.9. The decisive "proof" of near-infrared observations

In the early 90's a major progress was made in Astronomy, with the newly acquired capability of imaging galaxies in the near-infrared (especially in the K-band, at wavelengths close to 2μ). This relatively recent diagnostics is especially interesting because it allows us to probe the underlying older stellar component of galaxy disks (which accounts for most of the visible mass), while images in the optical are generally dominated by bright young stars and dust extinction. In practice, for a given object a comparison of images taken in the near-infrared and images taken in the optical allows us to disentangle the stellar component from the gaseous component of the disk.

Near-infrared observations have thus given the decisive "proof" to the Density Wave Theory. They have shown (starting with [34-36]) that large-scale spiral arms are a density perturbation in the stellar component of the disk, thus fulfilling a pledge made long ago by Oort (see Sect. 3.1). This demonstrates that, for such grand-design structure, other theories (such as magnetic theories, see [37], or the theory based on Stochastic Self-Propagating Star Formation [38]) are not viable. Furthermore, they have shown that grand design is very frequent, which suggests that only a theory based on intrinsic mechanisms can be reasonable. In addition, they have made it clear that such density perturbations in the underlying stellar disk are very smooth and sinusoidal, which encourages the direct application of a linear theory of global modes to the interpretation of the observed morphologies. Finally, by comparison with optical images, it has confirmed that grand design is generally two-armed while multiple-armed and less regular spiral structure is primarily a gaseous phenomenon.

3.10. Concluding remarks

In concluding this Section, I would like to comment on one frequently asked question: So, should one believe the *swing* theory or the *modal* theory of spiral structure? To me, this question is not well posed. In fact, swing [21] is primarily a mechanism. It is a very interesting mechanism in the dynamics of galaxy disks, but the picture advocated in the scenario (usually associated with the swing mechanism) in which spiral structure is transient and driven from the outside (see item (4) in Sect. 3.1) has not yet grown up into a complete theory. In addition, in the modal theory (presented in Sects. 3.5-3.7) the overreflection of leading into trailing waves, characteristic of the high-J regime described in Sect. 3.6, corresponds precisely to the swing mechanism, which is thus not ignored in modal studies.

The fact that a wide body of observations naturally fit in the framework of the Density Wave Theory grown into a theory of self-excited global modes (the "modal theory") is very encouraging and suggests that we have indeed reached one major step in understanding spiral structure in galaxies. A semi-empirical approach (in contrast to a deductive approach), by generating a large number of quantitative observational tests, has proved to be very successful and the winning approach for such complex systems as galaxies are. As a result, we now have a view of spiral galaxies in which the roles of gas and of dark halos are properly recognized. It would be very hard to justify the existence of a long-lasting, quasi-stationary spiral structure without the help of the dissipative, cold interstellar medium. Furthermore, without the help of the spheroidal bulge-halo component, fully self-gravitating disks would tend to generate only bars. The modal theory sets the problems of normal spiral structure and barred spiral structure within a unified framework. Near-infrared observations have shown that two-armed grand design spiral structure is basically ubiquitous, thus indicating that an explanation in terms of intrinsic processes is essentially unavoidable.

4 Other interesting problems in the dynamics of spiral galaxies and future prospects

Spiral galaxies offer a variety of interesting problems, many of which are not related, or only indirectly related, to the problem of spiral structure discussed in the main part of this paper. In particular, I should mention all the important issues that refer to the specific dynamical behavior of gas and of stars (dynamics of the interstellar medium, star formation processes, stellar orbits, etc.). Especially the cold interstellar medium, because of its clean radio signal at 21cm from atomic hydrogen, has often pointed to interesting dynamical problems, such as the origin of galaxy warps, for which a proper understanding would give us a better appreciation of the structure of galaxies, of their content of dark matter, and, in general, of their evolution. A short list of interesting research areas would thus include: the structure and dynamics of Low-Surface-Brightness galaxies, the presence and the properties of central massive black holes and their interactions with the host galaxies, the detailed distribution of dark matter in dark halos and some aspects of gravitational lensing, the Tully-Fisher relation and other scaling laws for spiral galaxies, problems of formation and evolution (also from the point of view of chemistry and stellar populations) as raised by observations of the distant universe at intermediate redshifts. From such an enormous range of possibilities, below I will spend only a few words on some specific topics that appear to be particularly exciting from the dynamical point of view.

4.1. Self-gravitating accretion disks

In Sect. 3.3 I had referred to the important mechanism of self-regulation in galaxy disks. In the last ten years several studies have addressed the dynamics of self-gravitating accretion disks, in a variety of contexts that range from the case of protostellar disks to that of Active Galactic Nuclei. In one line of research, focus has been placed on the role of self-regulation in such accretion disks (see [39], [26], and references therein). One prediction of these models (best applicable to the cold outer parts of protostellar and AGN accretion disks) is that self-regulated disks should be characterized by a flat rotation curve. There is at least one case (the nuclear disk of NGC 3079) where indeed such flat rotation curve appears to be present [40].

4.2. Extraplanar gas

Traditionally, the cold atomic hydrogen was thought to be confined to a very thin layer in the equatorial plane of spiral galaxies. Now it appears that some "anomalous" cold atomic hydrogen is often present well outside the equatorial plane, in a sort of slower rotating gaseous halo (for NGC 2403, see [41]; for NGC 4559, see [42]). This raises interesting modeling problems, related also to the detailed distribution of dark matter in these galaxies (for NGC 891, see [43]). Such cold gas could correspond to material ejected from the disk ("galactic fountains") or to fresh material still in the process of being slowly captured by the galaxy from its environment. Similar alternative scenarios are also involved in the classical problem of the interpretation of high velocity clouds in our Galaxy.

4.3. Hi disks in elliptical galaxies

Ellipticals were thought to be basically free from cold gas. Recent deep observations have shown that some ellipticals do possess regular disks of atomic hydrogen (for NGC 3108, see [44]). These might offer an excellent diagnostics for studies of dark matter (which, because of lack of straightforward kinematical tracers, have often reached controversial conclusions). In addition, one might study dynamical mechanisms in these disks and look there for spiral structure or other elements of continuity with the dynamics of spiral galaxies.

4.4. Prospects for new advances in the Density Wave Theory

The fact that the Density Wave Theory has reached a mature stage based on a very large number of quantitative contributions indicates that, at this point, significant advances in new directions will require major efforts. Obviously, in general it would be desired to investigate the properties of density waves and global modes in realistic three-dimensional multi-component models of galaxy disks, especially by extending the current knowledge to the non-linear level and to a full inclusion of stellar dynamical effects.

A list of specific topics, where concrete progress might be made within a realistic research program based on a fruitful exchange between Astrophysics and Applied Mathematics, is the following:

• Re-examine the shock scenario in the light of the new picture of the interstellar medium within the theory of global spiral modes.

• Re-examine the other classical observational tests of the Density Wave Theory in the light of the theory of global spiral modes and of the newly acquired observing capabilities.

• Study the problem of spiral structure in the gaseous outer disk and of its coupling with large-scale warps (for the case of NGC 6496, see [45]).

• Study the properties of damped global spiral modes and how some discrete modes of this type could be brought to be important by a suitable tidal interaction.

• Examine the issues of self-regulation and non-linear evolution, in view of statistical studies of spiral morphologies in the distant universe at intermediate redshifts.

Acknowledgments

I would like to thank the organizers of the Symposium for their invitation to this nice celebration in Beijing. My thanks also go to all the scientists that, together with C.C. Lin, have participated in the development of the Density Wave Theory.

References

1. C.C. Lin and F.H. Shu, *Astrophys. J.* **140**, 646 (1964)
2. C.C. Lin and F.H. Shu, *Proc. Nat. Acad. Sci.* **55**, 229 (1966)
3. Y.Y. Lau, C.C. Lin and J.W-K. Mark, *Proc. Nat. Acad. Sci.* **73**, 1379 (1976)
4. G. Bertin, Y.Y. Lau, C.C. Lin, J.W-K. Mark and L. Sugiyama, *Proc. Nat. Acad. Sci.* **74**, 4726 (1977)
5. D.J. Benney, F.H. Shu and C. Yuan, Eds., *Applied Mathematics, Fluid Mechanics, Astrophysics: A symposium to honor C.C. Lin*, World Scientific, Singapore (1988)
6. G. Bertin, in *IAU Symposium 100*, Ed. E. Athanassoula, Reidel, Dordrecht, p. 119 (1983)
7. G. Bertin, C.C. Lin and S. A. Lowe, in *Plasma Astrophysics ESA SP-207*, Eds. T.D. Guyenne and J.J. Hunt, ESA Scientific and Technical Publications, Noordwijk, p. 115 (1984)

8. G. Bertin, C.C. Lin, S.A. Lowe and R.P. Thurstans, *Astrophys. J.* **338**, 78 and 104 (1989)
9. G. Bertin and C.C. Lin, *Spiral Structure in Galaxies: A Density Wave Theory,* The MIT Press, Cambridge (1996)
10. G. Bertin, *Dynamics of Galaxies,* Cambridge University Press, Cambridge (2000)
11. A. Sandage, *The Hubble Atlas of Galaxies,* Publ. 618, Carnegie Institution, Washington (1961)
12. A. Sandage and G.A. Tammann, *A Revised Shapley-Ames Catalog of Bright Galaxies,* Publ. 635, Carnegie Institution, Washington, 2nd Ed. (1987)
13. A. Sandage and J. Bedke, *Atlas of Galaxies Useful for Measuring the Cosmological Distance Scale,* NASA SP-496, Washington (1988)
14. A. Sandage and J. Bedke, *The Carnegie Atlas of Galaxies,* Publ. 638, Carnegie Institution, Washington (1994)
15. E. Hubble, *Astrophys. J.* **64**, 321 (1926)
16. J.H. Oort, in *Interstellar Matter in Galaxies*, Ed. L. Woltjer, Benjamin, New York, p. 234 (1962)
17. W.W. Roberts, *Astrophys. J.* **158**, 123 (1969)
18. H.C.D. Visser, *Astron. Astrophys.* **88**, 149 and 159 (1980)
19. P.O. Lindblad, *Stockholm Observ. Ann.* **21**, 3 (1960)
20. P. Goldreich and D. Lynden-Bell, *Mon. Not. Roy. Astron. Soc.* **130**, 125 (1965)
21. A. Toomre, in *The Structure and Evolution of Normal Galaxies*, Eds. S.M. Fall and D. Lynden-Bell, Cambridge University Press, Cambridge, p. 111 (1981)
22. A. Toomre, *Astrophys. J.* **139**, 1217 (1964)
23. A. Toomre, *Astrophys. J.* **158**, 899 (1969)
24. C.C. Lin, in *IAU Symposium 38*, Eds. W. Becker and G. Contopoulos, Reidel, Dordrecht, p. 377 (1970)
25. G. Bertin and A.B. Romeo, *Astron. Astrophys.* **195**, 105 (1988)
26. G. Lodato and W.K.M. Rice, *Mon. Not. Roy. Astron. Soc.* **351**, 630 (2004)
27. J.W-K. Mark, in *IAU Symposium 58*, Ed. J.R. Shakeshaft, Reidel, Dordrecht, p. 417 (1974)
28. J.W-K. Mark, *Astrophys. J.* **205**, 363 (1976)
29. J.W-K. Mark, *Astrophys. J.* **212**, 645 (1977)
30. S. Kent, *Astron. J.* **93**, 1062 (1987)
31. A.J. Kalnajs, *Astrophys. Lett.* **11**, 41 (1972)
32. W.W. Roberts and F.H. Shu, *Astrophys. Lett.* **12**, 49 (1972)
33. G. Bertin, *Astron. Astrophys.* **127**, 145 (1983)
34. D.L. Block and R.J. Wainscoat, *Nature* **353**, 48 (1991)
35. D. Zaritsky, H.W. Rix and M.J. Rieke, *Nature* **364**, 313 (1993)
36. D.L. Block, G. Bertin, A. Stockton, P. Grosbøl, A.F.M. Moorwood and R.F. Peletier, *Astron. Astrophys.* **288**, 365 (1994)
37. J.H. Piddington, *Mon. Not. Roy. Astron. Soc.* **162**, 73 (1973)
38. H. Gerola and P.E. Seiden, *Astrophys. J.* **223**, 129 (1978)
39. G. Bertin, *Astrophys. J. Lett.* **478**, L71 (1997)
40. P.T. Kondratko, L.J. Greenhill and J.M. Moran, *Astrophys. J.* **618**, 618 (2005)
41. F. Fraternali, G. van Moorsel, R. Sancisi and T. Oosterloo, *Astron. J.* **123**, 3124 (2002)

42. C.V. Barbieri, F. Fraternali, T. Oosterloo, G. Bertin, R. Boomsma and R. Sancisi, *Astron. Astrophys.* **439**, 947 (2005)
43. M. Barnabè, L. Ciotti, F. Fraternali and R. Sancisi, *Astron. Astrophys.* **446**, 61 (2006)
44. T. Oosterloo, R. Morganti, E. Sadler, D. Vergani and N. Caldwell, *Astron. J.* **123**, 729 (2002)
45. R. Boomsma, T. van der Hulst, T. Oosterloo, F. Fraternali and R. Sancisi, in *IAU Symposium 217*, Eds. P.-A. Duc, J. Braine and E. Brinks, Astron. Soc. Pacific, San Francisco, p. 142 (2004)

DARK MATTER DYNAMICS IN GALAXIES

CHUNG-PEI MA
Department of Astronomy
University of California at Berkeley
Berkeley, CA 94720, USA
**E-mail: cpma@berkeley.edu*
http://astro.berkeley.edu/ cpma

During the first ten million years or so after the big bang, the fluctuations in matter and radiation in the universe remained small enough that they could be treated as tiny ripples imprinted on a smooth background. A major challenge in cosmology is in calculating how these small fluctuations grow under gravitational instability into highly collapsed objects. The traditional tool to study cosmological structure formation in the nonlinear region is numerical simulations. In this lecture I describe a complementary approach based on a kinetic theory for the evolution of the phase space distributions of dark matter in galaxy halos.

1. The New Cosmology

Cosmology, the study of the origin, evolution, and ultimate fate of the Universe, is perhaps at its most fascinating yet confusing stage of development. On the one hand, its framework — the big bang model — has withstood decades of tests and challenges. On the other hand, many basic facts about the Universe remain elusive: fundamental properties such as its energy and matter contents are among the most debated physical quantities today. Active observational and theoretical efforts in this field are currently leading to many interesting discoveries.

Among the most intriguing unsolved mysteries in cosmology is of what the Universe is made. Looking outward, we observe objects like planets, stars, and galaxies; looking inward, we detect particles like photons, electrons, and quarks. But is that all? Are we so privileged as to have witnessed the full spectrum of matter and radiation that Nature has to present? The advancement of science certainly has not left our egos unscathed. Four hundred years ago, after more than a thousand years of struggle, we were finally resigned to believe that the heavens did not rotate about the Earth. Eighty years ago, we were shown that there were many galaxies similar to our own spread out in a space so vast that even their light must travel millions of years to reach us. Just as we have come to accept that we live near an ordinary, middle-aged star in an ordinary galaxy occupying no special place in the cosmos, we are suddenly confronted with an even more humbling realization — the possibility that most of the mass in the Universe resides in "dark matter," some very clever form of matter capable of speeding up the motion of stars and galaxies while eluding direct detection at the same time!

Compelling evidence for the existence of dark matter is now abundant. In our solar system, more distant planets orbit with slower speeds as a result of the weaker gravitational pull of the Sun; by contrast, in a spiral galaxy, the stars and gas located farther away from the bulk of the light concentrated toward the center of the galaxy often show no trend of slowing down. Our perhaps too friendly neighbor, the Andromeda galaxy, is approaching us with a high speed that would defy Newton's law of gravity unless it is being helped secretly by a large quantity of invisible matter. The galaxies in clusters — aggregations of hundreds to thousands of galaxies held together by gravity — are swarming around so fast that the luminous matter can make up only less than 1% of the total mass. These and other observations all point toward a single fact: more mass is out there than we can see. More startlingly, as we look on grander scales, Nature's ability to hide dark matter increases.

2. Pros and Cons of Numerical Simulations

The standard tool for studying the nonlinear growth of cosmic structures is numerical simulation. Numerical simulation is a kind of computer experiment particularly powerful for studying complicated problems that cannot be explored in a laboratory setting and can only be solved to a limited extent with analytical methods. The study of the growth of structure and clustering of galaxies is one such example. The advent of modern supercomputers has made it possible for us to simulate these cosmic processes that take place over billions of years in a representative patch of the Universe millions of light-years across. Typically we take the last analytical answer we could obtain and trust from methods such as the linear perturbation theory, and use it as the input for the subsequent simulations on supercomputers. Millions to billions of fictitious particles are used to represent the motion of the dark matter, which is governed by the law of gravity. The computer is instructed to calculate the force on each particle due to all other particles at a given time. It then advances the position and updates the velocity of every particle over a time interval that is short compared to the rate of change of the forces. This process is typically repeated thousands of times until the present cosmic time is reached.

Despite the power of simulations, it is important to keep in mind their limitations. For one, the particle masses used in even the highest resolution cosmological N-body simulations today are at least *60 orders of magnitude* larger than the masses of individual cold dark matter (CDM) particles. The CDM phase space is therefore sampled very coarsely in cosmological N-body simulations. In addition to these numerical artifacts arising from limited dynamic range or artificial two-body relaxation, there are some problems, e.g. the massive black hole merger problem, where the numerical artifacts seriously limit the ability of the N-body method to simulate faithfully the physics. Another drawback of numerical simulations is that they offer no direct analytical insight into the outputs. We typically gain insight by performing a suite of simulations varying the parameters, but an analytical approach is preferable if it can isolate and properly describe the essential physics.

Motivated by these issues with simulations, we have begun to explore an alternative approach in a recently paper [1] in which we use the kinetic theory to describe the evolution of the phase-space distribution of dark matter particles in galaxy halos in the presence of a cosmological spectrum of fluctuations. This theory introduces a new way to model the formation and evolution of dark matter halos, which traditionally have been investigated by analytic gravitational infall models or numerical N-body methods. Further development of this theory and its potential applications are underway. This kinetic description should provide a framework for understanding the results of numerical simulations and for guiding further research into the physics of dark matter. In current hierarchical models, galaxy halos grow by both frequent minor mergers (or accretion) of smaller mass halos and occasional major mergers with another halo of comparable or larger mass. Traditional N-body simulations are still the method of choice for studying major mergers, whereas the kinetic approach described below should provide a good description for the effects of frequent minor mergers.

The dynamics of globular clusters is an interesting case for comparison. This is one astrophysical system that has been well studied with both N-body simulations (e.g. Ref. 2) and kinetic theories based on the Fokker-Planck equation (e.g. Ref. 3, 4, 5, 6). The N-body technique is particularly suitable for globular clusters because the number of stars in a globular is $\sim 10^5$ to 10^6, which is comparable to the number of simulation particles that can be accommodated in modern computers and special-purpose hardware. However, much of our understanding of the basic physics of cluster evolution, such as core collapse, has first come from kinetic theory. Just as in globular cluster studies, we will use N-body simulations to test and calibrate our kinetic theory as necessary, and as a numerical laboratory for exploring new effects revealed by the kinetic theory.

3. Cosmological Kinetic Theory

Kinetic theories are generally used to describe the evolution of the distribution function or phase space density, $f(\vec{r}, \vec{v}, t)$. It is normalized so that $\rho(\vec{r}, t) = \int f\, d^3v$ is the mass density of particles at position $\vec{r}$ and time t. For an ideal classical gas, f follows a Maxwellian distribution in velocity $\vec{v}$ with mean velocity, temperature, and net mass density that may be functions of position and time. For dark matter, however, the velocity distribution generally is non-Maxwellian, and numerical simulation or kinetic theory must be used to determine f before ρ can be calculated. Our approach focuses on the phase space density as the key to understanding dark matter halos.

3.1. *A sketch of the derivation*

The starting point of our work [1] is a rigorous derivation of a kinetic equation for dark matter evolution in second-order cosmological perturbation theory. We begin with the one-particle phase space density for dark matter particles

$$f_{\rm K}(\vec{r},\vec{v},t) = m\sum_i \delta_{\rm D}[\vec{r}-\vec{r}_i(t)]\delta_{\rm D}[\vec{v}-\vec{v}_i(t)]\,, \tag{1}$$

where δ_D is the Dirac delta function, and its evolution equation

$$\frac{\partial f_{\rm K}}{\partial t} + \vec{v}\cdot\frac{\partial f_{\rm K}}{\partial \vec{r}} + \vec{g}_{\rm K}\cdot\frac{\partial f_{\rm K}}{\partial \vec{v}} = 0\,, \quad \vec{g}_{\rm K}(\vec{r},t) = -G\int d^6w'\, f_{\rm K}(\vec{w}\,',t)\frac{(\vec{r}-\vec{r}\,')}{|\vec{r}-\vec{r}\,'|^3}\,, \tag{2}$$

where we have grouped all six phase-space variables into $\vec{w}$ for notational convenience. Rather than giving a perfect description of a single halo, we average over halos to obtain a statistical description of halo evolution:

$$f(\vec{w},t) \equiv \langle f_{\rm K}(\vec{w},t)\rangle\ . \tag{3}$$

The third term in Eq. (2) depends on the product of two $f_{\rm K}$, which can be written as

$$\langle f_{\rm K}(\vec{w}_1,t)f_{\rm K}(\vec{w}_2,t)\rangle \equiv \delta_{\rm D}(\vec{w}_1-\vec{w}_2)f(\vec{w}_1,t) + f(\vec{w}_1,t)f(\vec{w}_2,t) + f_{2c}(\vec{w}_1,\vec{w}_2,t)\,, \tag{4}$$

where f_{2c} is the two-point correlation function in phase space. We then obtain the evolution equation for an average halo:

$$\frac{\partial f}{\partial t} + \vec{v}\cdot\frac{\partial f}{\partial \vec{r}} + \vec{g}_{\rm T}\cdot\frac{\partial f}{\partial \vec{v}} = -\frac{\partial}{\partial \vec{v}}\cdot\vec{F}_v\ , \tag{5}$$

where

$$\vec{F}_v(\vec{x},\vec{v},t) = -G\int d^3x'd^3v'\, f_{2c}(\vec{x},\vec{v},\vec{x}\,',\vec{v}\,',t)\left(\frac{\vec{x}-\vec{x}\,'}{|\vec{x}-\vec{x}\,'|^3} + \frac{\vec{x}\,'}{|\vec{x}\,'|^3}\right)\ . \tag{6}$$

Here, $\vec{g}_{\rm T} = \vec{g}(\vec{r},t) - \vec{g}(0,t)$ is the gravitational tidal field where $\vec{g}$ is the gravity field produced by $\rho(\vec{r},t)$. We have subtracted out $\vec{g}(0,t)$ because only the tidal field is relevant to halo structure and evolution.

The right-hand side of Eq. (5) is the gravitational tidal acceleration per unit volume arising from two-point correlations of particles in phase space; f_{2c} is the phase-space two-point correlation function, a generalization of the well-known two-point correlation $\xi(r)$ for matter clustering. This term arises because we have taken an ensemble average over halos in order to describe statistically the substructure within halos. Heuristically, f_{2c} describes the substructure within a galaxy halo at the two-point level; higher order correlation functions would be needed for a complete description. For example, the initial density field has fluctuations that are progenitors of the many small halos that form and merge hierarchically later. The lumpiness of the matter distribution represents a fluctuation about the average (spherical) density field. These fluctuations cause changes in the energy and angular momentum of individual particle orbits that are crucial to the actual evolution.

Eq. (5) is the first BBGKY hierarchy equation. It is incomplete because it does not give an expression for the phase space two-point correlation f_{2c}. In Ref. 1, we were able to evaluate f_{2c} exactly in second-order cosmological perturbation theory using the BBKS formulation [7] of the statistical properties of constrained Gaussian density peaks. We obtained an expression for the right-hand side of Eq. (5) in the quasilinear regime, which has the general form

$$\vec{F}_v = \vec{\alpha} f - \gamma \vec{v} f - \mathbf{D} \cdot \frac{\partial f}{\partial \vec{v}} . \tag{7}$$

This is precisely the form of a Fokker-Planck flux [5,6] and has three transport coefficients: *drift* $\vec{\alpha}$, *drag* γ, and *diffusivity* $\mathbf{D}$. In general, all of these coefficients can be functions of $(\vec{r}, \vec{v}, t)$. For a spherical average halo, $\vec{\alpha} = \alpha \hat{r}$ is radial. The well-known dynamical friction is described by $\gamma(v)$. Specifically, we obtained the following results for the transport coefficients in second-order cosmological perturbation theory:

$$\vec{\alpha}(\vec{r}, t) = \mathrm{Cov}(\delta, \vec{g}_{\mathrm{T}} | \vec{v}\,) \ , \quad \gamma = 0 \ , \quad \mathbf{D}(\vec{r}, t) = \mathrm{Cov}(\vec{g}_{\mathrm{T}}, \vec{v}\,) \ , \tag{8}$$

where $\delta = \delta\rho/\rho$ is the density perturbation, and Cov denotes the covariance over cosmological random fields defined by

$$\mathrm{Cov}[A, B] \equiv < (A - < A >)(B - < B >) > = < AB > - < A > < B > \ . \tag{9}$$

Explicit expressions for $\vec{\alpha}$ and $\mathbf{D}$ as integrals over the power spectrum $P(k)$ of cosmological density perturbations are given in Eqs. (28)-(30) of Ref 1.

3.2. *Subtleties*

Traditional derivations of the Fokker-Planck equation are based on the Master Equation, a phenomenological equation that assumes that the dynamics be a Markov process, a strong assumption that is not always valid. We emphasize that we did not follow this approach. Instead, our derivation leading to the Fokker-Planck equation started with the first BBGKY hierarchy equation and is *exact* to second order in cosmological perturbation theory. We have shown that the force fluctuations arising from substructure lead to dissipation, and that a full N-body treatment is not necessary to describe this dissipation for the average halo. The Fokker-Planck equation describes the evolution of weakly collisional gas and characterizes the slow relaxation mechanisms that drive a system towards equilibrium.

Eq. (8) has several surprises. First, we found that to second order in perturbation theory there is no dynamical friction: $\gamma = 0$. Instead there is a radial drift $\alpha(r, t)$, a term unfamiliar to astrophysicists. It arises from the clustering of substructure within a halo. Sub-halos interior to a given radius r are correlated with density fluctuations at r, leading to a correlated force density that is not described by the average density profile. We showed that models with much small-scale power and substructure ($n > -2$ as $k \to \infty$, where the matter fluctuation power spectrum is

$P(k) \propto k^n$) have a strong inward drift force, while models that are smoother on small scales ($n < -2$) have vanishing drift force as $r \to 0$.

We also found that the eigenvalues of the diffusivity tensor $\mathbf{D}$ can be *negative.* Negative diffusivity causes the velocity dispersion (or temperature) to decrease and leads to a thermodynamic instability. The cause is the enhancement of gravitational instability by second-order perturbations [8]. In the strongly nonlinear regime, after virialization, we expect the diffusivities to become positive.

Finally, we found that the initial relaxation timescale due to drift and diffusion is comparable to the Hubble time. This means that relaxation processes due to substructure are significant during the initial stages of halo formation when the Hubble time was short. Drift and diffusion will significantly modify the evolution of the average halo compared with the idealized spherical infall solutions of Ref. 9 and 10. We now have a framework in which to compute these correction effects.

3.3. *Extension into the nonlinear regime*

Our derivation of Eq. (8) is fully analytical and the resulting equation describes the early phase of halo evolution, but it was valid only to second order in cosmological perturbation theory. We suspect that the result did not yield a dynamical friction term (i.e. $\gamma = 0$) because our calculation was limited to small-amplitude perturbations about a homogeneous and isotropic expanding cosmological model. In the fully nonlinear regime the drift and diffusivity will certainly be modified and we expect dynamical friction to appear. In particular, we expect two types of drift terms to be present: $\vec{A} = \alpha\hat{r} - \gamma\vec{v}$, where α is the radial drift and γ is the drag coefficient. The Chandrasekhar calculation [11] suggests that the drag and its accompanying diffusivity will depend on both position (through ρ_b) and velocity.

We conjecture that the Fokker-Planck description is approximately valid when the matter distribution is modeled as a set of clumps (i.e., the halo model) that scatter individual dark matter particles away from the orbits they would have in a smooth, spherical potential. As a first step toward understanding the effects of substructures on the dark matter phase-space distribution in the nonlinear regime, we have performed a series of fully dynamical numerical simulations to study the gravitational interplay between a host halo and its subhalos in a controlled and semi-realistic way [12]. We used subhalo properties similar to those found in earlier full-scale cosmological simulations [13,14] and placed roughly 10% of a host halo's mass in the form of a thousand smaller, dense satellite subhalos with a subhalo mass function $dn_{sub}/dM_{sub} \propto M_{sub}^{-\alpha}$, where $\alpha \sim 1.7 - 1.9$. This approach allowed us to perform a suite of numerical experiments to quantify the effects due to a wider range of subhalo masses, concentration, and orbits than was possible with large cosmological simulations.

Depending on the competition between the addition from subhalo masses deposited in the central regions and the removal of main halo particles due to gravitational heating, we find that the inner cusp of the *total* mass density can steepen, remain the same, or flatten [12]. For instance, in a model where the total subhalo mass is 7% of the host halo mass and the most massive two subhalos have 1.51% and 1.25% M_{host}, the subhalos suffer much tidal mass losses and do not add much mass to the central part of the halo. As a result, we found the inner density profiles of both the host halo and the sub+host halo to flatten from the initial $\rho \sim r^{-1}$ to $r^{-0.75}$ in ~ 6 dynamical times. In contrast, in a model with 10.3% subhalo mass and concentration parameter $c_{sub} = 31.2$, the mass added by the most massive subhalos (the top two have 4.66% and 2.09% M_{host}) more than compensate for the flattening in the host halo, leading to a steeper than r^{-1} inner cusp.

This numerical study of the nonlinear regime suggests that fluctuations due to subhalos in parent halos are important for understanding the time evolution of dark matter density profiles and the halo-to-halo scatter of the inner cusp seen in recent ultra-high resolution cosmological simulations [15]. We have shown that this scatter may be explained by subhalo accretion histories: when we allow for a population of subhalos of varying concentration and mass, the total inner profile of dark matter can either steepen or flatten.

Extending our derivation of the second-order cosmological kinetic equation discussed in the earlier part of this lecture into the non-linear regime will provide further insight into the diffusion effects on dark matter halos due to substructures seen in our numerical experiments.

4. Acknowledgments

The kinetic theory described in this lecture is developed in collaboration with Ed Bertschinger. CPM is supported in part by NASA grant NAG5-12173 and NSF grant AST 0407351. The research used resources of the National Energy Research scientific Computing Center, which is supported by the Office of Science of the U.S. Department of Energy under Contract No. DE-AC03-76SF00098.

References

1. Ma, C.-P. & Bertschinger, E. 2004, ApJ, 612, 28
2. Heggie, D. & Hut, P. 2003, The Gravitational Million-Body Problem: A Multidisciplinary Approach to Star Cluster Dynamics (Cambridge: Cambridge University Press)
3. Fokker, A. D. 1914, Ann. Physik 43, 810
4. Planck, M. 1917, Sitzber. Preuf. Akad. Wiss. p. 324
5. Spitzer, L. 1987, Dynamical evolution of globular clusters (Princeton: Princeton University Press)
6. Binney, J.J. & Tremaine, S. 1988, Galactic Dynamics (Princeton: Princeton University Press)
7. Bardeen, J.M., Bond, J.R., Kaiser, N., & Szalay, A. S. 1986, ApJ, 304, 15
8. Peebles, P.J.E. 1980, The Large Scale Structure of the Universe (Princeton: Princeton University Press)
9. Fillmore, J. A. & Goldreich, P. 1984, ApJ, 281, 1
10. Bertschinger, E. 1985, ApJS, 58, 39
11. Chandrasekhar, S. 1943, ApJ, 97, 255

12. Ma, C.-P. & Boylan-Kolchin, M. 2004, Phys. Rev. Lett. 93, 021301
13. Klypin, A. et al. 1999, ApJ, 522, 82
14. Ghigna, S. et al. 2000, ApJ, 544, 616
15. Navarro, J. et al. 2004, MNRAS, 349, 1039; Diemand, J., Moore, B. & Stadel, J. 2004, MNRAS,353, 624; Reed, D. et al. 2005, MNRAS, 357, 82

ASYMPTOTICS AND STAR FORMATION

FRANK H. SHU†

Physics Department, University of California at San Diego, 9500 Gilman Drive, SERF 408
La Jolla, CA 92093-0424, USA

Star formation is a process that spans many decades in length and time scales. As a consequence, asymptotic methods of the variety pioneered by C. C. Lin and other applied mathematicians can be used to great effect at various stages of the problem. In this lecture, we give an overview of the complete problem in the case of the formation of single stars of sunlike masses, and we highlight a few of the mathematical solutions made possible by an application of asymptotic techniques and ideas – the formation of molecular cloud cores; the gravitational collapse of self-similar, magnetized, rotating toroids; the dissipation by magnetic reconnection of the trapped interstellar flux brought in by gravitational collapse that would otherwise have prevented the formation of circumstellar disks; and the jets and outflows that result when the accreting circumstellar disk interacts with the magnetosphere of the newly formed star. We end by indicating where additional progress requires a better understanding of the role of turbulence, another subject where astronomers of the future can look to the work of C. C. Lin for guidance.

1 Introduction

I am very pleased to be able to participate in the celebration of Professor C. C. Lin's 90th birthday. I note that an interesting commensurability occurs this year in his age (90) and of Tsinghua University in Beijing (95) and Tsing Hua University in Hsinchu (50). The hotel where the participants of this Symposium are staying is called the "Purple Splendor" in Chinese. I think of it as a metaphor for C. C. Lin returning to his alma mater Tsinghua (whose school color is purple). It is to be hoped that this special conjunction will bring forth many new young stars.

1.1. *Efficacy of Asymptotic in Astrophysics*

Among the many things mathematical I learned from C. C. Lin, the most important has probably been asymptotics, which has a special efficacy in theoretical astrophysics. Astronomy can be said to be the science of large and small numbers. When large or small dimensionless parameters appear in a problem, opportunity for a parameter expansion often presents itself. Many astronomical problems have a large dynamic range

† Formerly at National Tsing Hua University, Hsinchu 30013, Taiwan, ROC, where my work was supported by a National Science Council grant to the Theoretical Institute for Advanced Research in Astrophysics (TIARA).

in space or time; as a consequence, similarity methods can also frequently be used to good effect.

1.2. *Four Phases of Star Formation Whether Isolated or Clustered*

In this lecture I shall give four examples from the field of star formation. Figure 1 depicts schematically the four phases of the formation of a single sunlike star, as they have been deduced from a combination of theory and observation (Shu, Adams, & Lizano 1987). We begin with phase (a), the quasi-static gravitational condensation of dense cores from a giant molecular cloud through the dissipation of turbulence and the slippage of neutrals past ions and the magnetic fields to which they are attached. The slippage occurs by a process of ambipolar diffusion first described by Mestel and Spitzer (1956). Phase (a) is deduced empirically to last between 1 to 3 million years (Myr). When the central regions of a core become sufficiently dense, the core collapses dynamically from "inside-out" and passes to phase (b). If the core possesses even a small amount of rotation, the gas and dust do not fall directly into the center, but instead swirl into a centrifugally supported ("Keplerian") disk, which slowly accretes onto the central star by an incompletely understood "viscous" process of transfer of angular momentum outwards and mass inwards. The duration of phase (b) is not known; sources undergoing

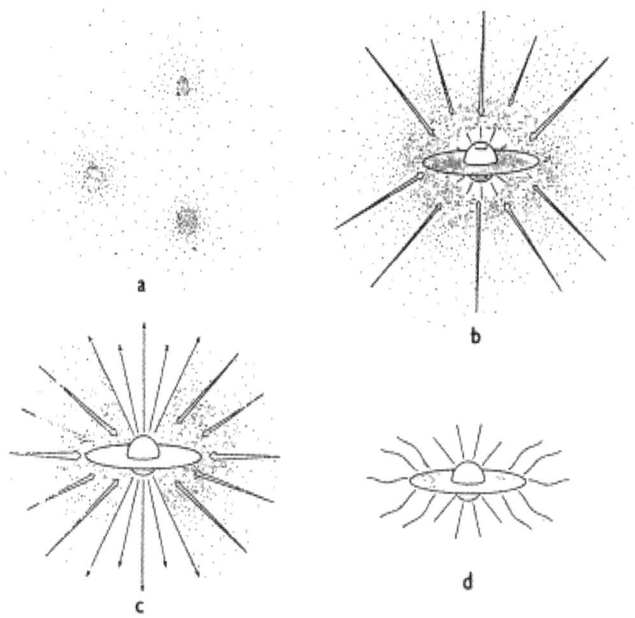

Figure 1. The four stages of star formation. (a) Cores form within molecular clouds as ambipolar diffusion expels magnetic support and interstellar turbulence decays. (b) A protostar with a surrounding nebular disk forms at the center of a cloud core collapsing from inside-out. (c) A stellar wind breaks out along the rotational axis of the system, creating a bipolar outflow. (d) The infall terminates, revealing a newly formed star with a circumstellar disk out of which a planetary system may be born. (From Shu, Adams, & Lizano 1987.)

pure (rotating) infall have not yet been found empirically. Thus, since hundreds of "starless cores" in phase (a) have been examined observationally (Jajina, Adams, & Myers 1999), phase (b), if it exists at all, must last no longer than perhaps ~ 0.01 Myr. After that time, the system apparently makes a transition to a bipolar outflow phase (c), which we can estimate empirically, on the basis of the ratio of cores with stars that have outflows to starless cores, to last between 0.1 to 0.4 Myr. In phase (c), outflow occurs in two diametrically opposed directions along the rotation axis of the system while gravitational infall continues in the equatorial regions. Over time, the outflow angle gradually widens, until the system makes a transition to phase (d), when the inflow is completely reversed by the outflow over 4π steradians, except perhaps for a narrow range of angles encompassing the equatorial disk where a slow, viscously driven, accretion still takes place. Sunlike objects in phase (d) are called T Tauri stars, and they are seen by outside observers as a warm visible object at the center surrounded by a cooler flattened disk emitting copiously in the infrared

1.3. *Equations of Non-ideal Magnetohydrodynamics*

The equations governing the behavior of the material in all four phases of Fig. 1 are those of non-ideal magnetohydrodynamics (MHD):

$$\frac{\partial \rho}{\partial t} + \nabla \cdot (\rho \vec{u}) = 0, \tag{1}$$

$$\frac{\partial \vec{u}}{\partial t} + \nabla\left(\frac{1}{2}u^2\right) + (\nabla \times \vec{u}) \times \vec{u} = -\nabla U - \frac{1}{\rho}\nabla P + \frac{1}{4\pi\rho}\left(\nabla \times \vec{B}\right) \times \vec{B} + \frac{1}{\rho}\nabla \cdot \vec{\pi}, \tag{2}$$

$$\nabla^2 U = 4\pi G \rho, \tag{3}$$

$$\frac{\partial \vec{B}}{\partial t} + \nabla \times (\vec{B} \times \vec{u}) = \nabla \times \left(-\eta \nabla \times \vec{B} - \kappa \vec{B} \times (\nabla \times \vec{B}) + \frac{\tau \vec{B}}{4\pi\rho} \times \left[\vec{B} \times (\nabla \times \vec{B})\right]\right), \tag{4}$$

where $\rho, \vec{u}$, and $\vec{B}$ are, respectively, the volume density, fluid velocity, and magnetic field. In Eq. (2), $\vec{\pi}$ is the viscous stress tensor and is assumed to be given in terms of the deformation-rate tensor $\vec{\vec{D}}$ (traceless, symmetric, rate of strain) via the Newtonian relation, $\vec{\vec{\pi}} = \mu \vec{\vec{D}}$, where μ is the coefficient of shear viscosity. In the perfect gas law, $P = \rho kT / m$, where k is Boltzmann's constant and m is the mean molecular mass, the temperature T is set, in principle, by radiative processes. For many of the stages that we are concerned with in this review, the combination, $a^2 = kT / m$, which is the square of the isothermal speed of sound, may be approximated to be a constant.

In Eqs. (2) and (3), U is the gravitational potential of the system, while G is the universal gravitational constant. Equation (4) represents Faraday's law of induction in an electrically conducting but lightly ionized medium, where η is the electrical resistivity, κ is the Hall coefficient, and τ is the collision time between a typical neutral molecule and the ions of the medium. The transport coefficients η, κ, and τ are given by microscopic collision processes among ions, electrons, and neutrals, and they are greatly enhanced in dusty regions shielded from ultraviolet radiation because they depend inversely on the ionization fraction, which is very low in dark molecular-cloud cores and the interior layers of protoplanetary disks. The approximation of ideal MHD results if we set μ, η, κ, and τ all equal to zero.

2 Molecular Cloud Cores

2.1. *Gravitational Condensation by Ambipolar* Diffusion

In our first application we ignore all diffusive effects except for ambipolar diffusion; i.e,, we set $\mu = \eta = \kappa = 0$, with $\tau \neq 0$ having a functional form appropriate to the low-ionization conditions of molecular cloud cores. Figure 2 shows the resulting numerical

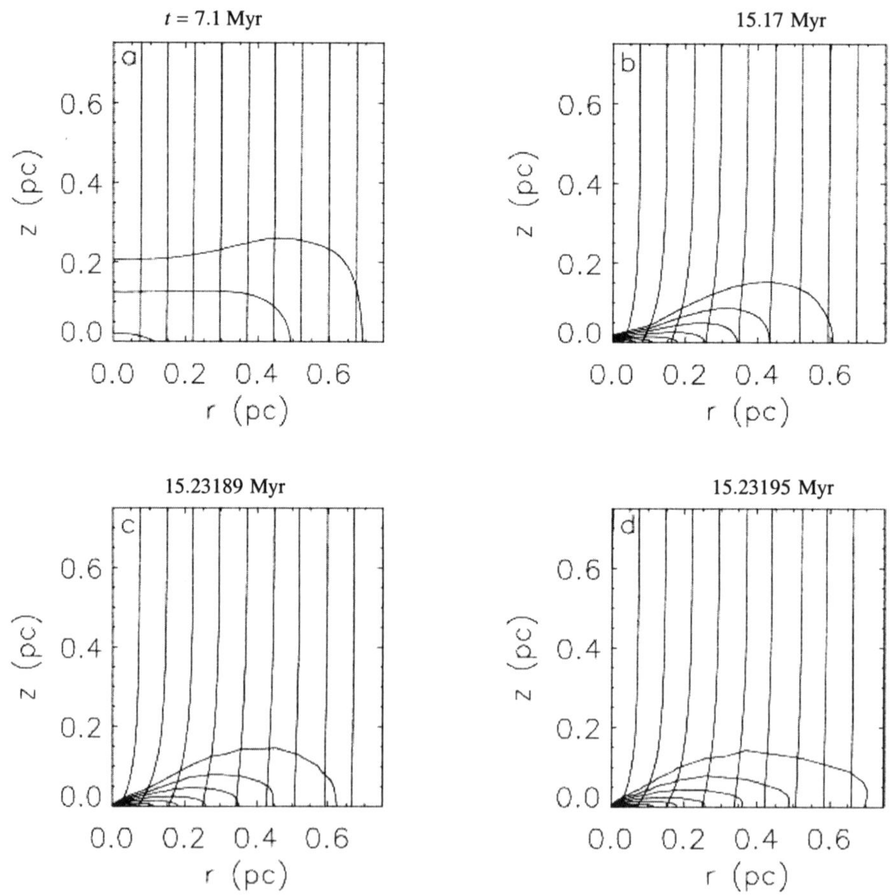

Figure 2. The formation of a molecular cloud core by laminar ambipolar diffusion. (From Desch & Mouschovias 2001.)

calculation by Desch and Mouschovias (2001) assuming axial symmetry and adopting an approximation that the configuration is in quasi-static equilibrium along the vertical direction in which it is highly flattened. The molecular cloud core (see isodensity contours given by solid curves) condenses across and along nearly vertical field lines that are slightly pinched inwards because neutrals drag on the ions as the neutrals contract under the action of their self-gravity. Because the condensation speeds are generally small compared to the isothermal sound speed a, quasi-static equilibrium is a good approximation also in the cylindrically radial direction until the very end. The passage of model time, beginning with an arbitrary zero for a not very condensed "initial state," is marked at the top of each panel. Notice that the displayed intervals become shorter and shorter as the central concentration in the isothermal approximation approaches a formally infinite value just past the last panel. Although specific to a certain model, the general features – in particular, the monotonic and catastrophic increase of the central density compared to the surroundings on a time scale $\sim 10^7$ yr – are characteristic of all laminar numerical simulations to date (e.g., Nakano 1979, Lizano & Shu 1989).

If one compares the duration predicted between when one has a recognizable cloud core (say, central density sufficient to excite the ammonia molecule into measurable emission, Myers & Benson 1983) and when a near-infinite central density is reached with observations (beyond which one has an accreting protostar at the center; see below), one arrives at a discrepancy. The statistics of cores with stars and without stars imply that the observed duration is shorter than the computed laminar models by a factor of 3 to 10 (Jajina, Myers, & Adams 1999).

Myers & Lazarian (1998) have proposed that turbulent support and its decay are responsible for the shorter time scales needed to condense molecular cloud cores, whereas Zweibel (2002) and Fatuzzo & Adams (2002) suggest that turbulent mixing enhances the rate of ambipolar diffusion. Two-dimensional simulations performed by Li &Nakamura (2003) suggest that a combination of turbulence and ambipolar diffusion can indeed reduce core-formation time-scales to the range 1-3 Myr mentioned in Section 1.2.

2.2. *Pivotal State*

Circumstantial evidence exists therefore that the set of nonlinear partial differential equations governing the formation of molecular cloud cores by ambipolar diffusion have *gravomagneto catastrophe* as an asymptotic attractor state. Independent of starting conditions, as long as the system is gravitationally bound and not too far from mechanical equilibrium, the center tries to acquire a singular condition of infinite density in finite time. For sake of definiteness, let us reset time to zero at the moment of the catastrophe; thus, $t < 0$ represents the stage of slow contraction leading up to the catastrophe, phase (a) in Fig. 1; whereas $t > 0$ represents the stage of dynamical collapse that follows, phase (b) in Fig. 1 leading to the later evolutionary stages of bipolar-outflow phase (c) and T Tauri phase (d). Let us also adopt the nomenclature of Li & Shu (1996) in calling the configuration at the transitional instant $t = 0$, the pivotal state.

Although it might be possible to demonstrate semi-analytically the convergence of all solutions, within limits, to a unique pivotal state at $t = 0$, no one has yet given an explicit proof. What has been done is to guess that the axisymmetric, pivotal state, written in spherical polar coordinates (r, θ, φ), is quasi-static and has radial dependences for the mass density, magnetic field strength, and angular velocity that satisfy $\rho \propto 1/r^2, B \propto 1/r$, and $\Omega \propto 1/r$. In these circumstances, for the non-rotating case, we suppose that the density and flux function (essentially, the vector potential which can be taken to lie entirely in the φ direction):

$$\rho(r,\theta) = \frac{a^2}{2\pi G r^2} R(\theta), \tag{5}$$

$$\Phi(r,\theta) = \frac{4\pi a^2 r}{G^{1/2}} \phi(\theta), \tag{6}$$

where $R(\theta)$ and $\phi(\theta)$ are functions to be determined. A linear sequence of models can be obtained, with the position in the sequence being determined by the parameter H_0 that measures the over-density supported against self-gravity by the magnetic field in comparison with what can be supported by thermal pressure alone:

$$\int_0^{\pi/2} R(\theta)\sin\theta d\theta = 1 + H_0. \tag{7}$$

The equations of magnetostatic equilibrium then reduce to the following set of ordinary differential equations for the functions $R(\theta)$ and $\phi(\theta)$:

$$\frac{1}{\sin\theta}\frac{d}{d\theta}\left[\sin\theta\left(2H_0\frac{\phi'}{\phi} - \frac{R'}{R}\right)\right] = 2\left(R - 1 - H_0\right), \tag{8}$$

$$\phi\frac{d}{d\theta}\left(\frac{\phi'}{\sin\theta}\right) = -H_0 R\sin\theta, \tag{9}$$

which are to be solved subject to the boundary conditions: zero flux at the pole, $\phi = 0$ at $\theta = 0$, and $2H_0\phi'/\phi - R'/R = 0$ at $\theta = \pi/2$. The latter is required so that Eq. (8) when multiplied by $\sin\theta$ and integrated from $\theta = 0$ to $\theta = \pi/2$ is consistent with Eq. (7). Numerical integrations then yield the results of Fig. 3, where contours of constant ρ and Φ (which label field lines) are plotted in the meridional plane for a selection of values of the control parameter H_0. These semi-analytic, self-similar, isothermal toroids, can be favorably compared to the end state of the numerical simulation of Fig. 2 if we choose a value for H_0 approximately equal to 1.*

The degree which a cloud core can approach an infinite central concentration before it goes into dynamical collapse is controversial, both from the theoretical and observational points of view. In a certain sense, the question is one of semantics. How small does the contraction speed u need to be in comparison to the sound speed a in order to qualify for the description "quasistatic?" If one arbitrarily chooses $u = a/2$, then theory and observations agree that the densities of the centers of molecular cloud cores exceed envelope values by several orders of magnitude.

---

*My usage of the word "semi-analytic" means something precise, the reduction of partial differential equations to a set of ordinary differential equations – or, at worst, a set of integro-differential equations in a single variable. In an age of electronic computation, the solution is then almost equivalent to classical functions called "analytic" by previous generations of scientists where someone else has taken the trouble to tabulate the dependences of such functions on their arguments and dimensionless parameters.

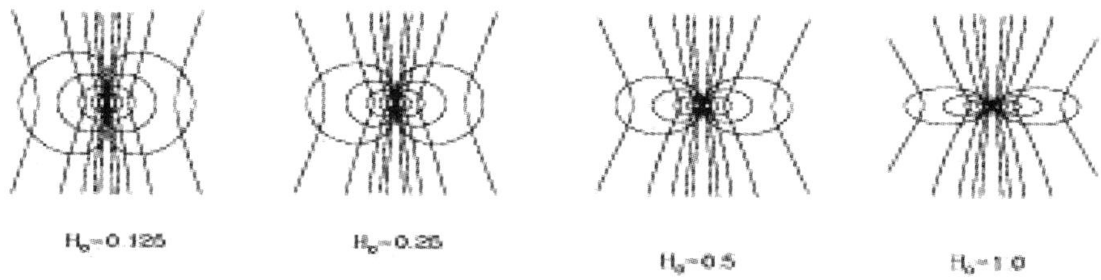

Figure 3. Isodensity contours and field lines spaced at logarithmic intervals in the meridional plane for self-similar, non-rotating, singular isothermal toroids. (From Li & Shu 1996.)

3 Gravitational Collapse of Self-similar Pivotal States

If we assume that the pivotal state at $t = 0$ is given by one of the static, singular configurations represented by Fig. 3, then the subsequent dynamical collapse is self-similar in space and time. The reason is that no characteristic length or time scale exists in the problem if the gravomagneto catastrophe process produces pure power-law distributions of density and field. The only dimensional parameters of the problem are the universal gravitational constant G and the isothermal sound speed a from which one cannot derive a characteristic length or time. In the subsequent evolution, reduced nondimensional variables that are scaled properly can depend on the coordinate and time variables (r,θ,t), only in the combination $\xi \equiv r/at$ and θ. In other words, for given H_0, the solution for $t > 0$ must take the form:

$$\rho(r,\theta,t) = \frac{\alpha(\xi,\theta)}{4\pi G t^2}, \tag{10}$$

$$\vec{u}(r,\theta,t) = a\vec{\mathrm{v}}(\xi,\theta), \tag{11}$$

$$\Phi(r,\theta,t) = \frac{4\pi a^3 t}{G^{1/2}}\Psi(\xi,\theta), \tag{12}$$

with reduced variables $\alpha, \vec{\mathrm{v}}$, and Ψ having the asymptotic forms,

$$\alpha(\xi,\theta) \to \frac{2}{\xi^2}R(\theta),\ \Psi(\xi,\theta) \to \xi\phi(\theta),\ \vec{\mathrm{v}}(\xi,\theta) \to 0,\ \text{ as } \xi \to \infty, \tag{13}$$

that is appropriate to the pivotal state at $t = 0$.

In practice, the resulting partial differential equations for the reduced variables α, v, and Ψ are still not solvable analytically for finite ξ and θ; thus, it is more convenient, because of the existence of standard packages in numerical MHD, to modify such programs slightly to do the simulations in the full non-reduced variables. Figure 4 gives the result of one such simulation for the case $H_0 = 0.25$ (Allen, Shu, & Li 2003). The main difference with the inside-out collapse solution known from semi-analytic studies of the non-rotating, spherical problem $H_0 = 0$ (Shu 1977) is the formation of a structure flattened by the anisotropic magnetic forces called a *pseudodisk*. Otherwise, the most important result from the numerical work is the derivation of the rate of mass accumulation by a growing protostar at the center,

$$\dot{M} = m_0(1+H_0)\frac{a^3}{G}, \tag{14}$$

where, for all values of H_0 from zero to infinity, m_0 is a numerical coefficient lying within 5% of the value 0.975 appropriate to $H_0 = 0$. For $a = 0.2$ km/s and $H_0 = 1$, equation (14) yields $\dot{M} = 3M_{\odot}$ /Myr, implying that it takes 0.17 Myr to form a half solar-mass star, which is the typical outcome of low-mass star-formation.

Although we cannot obtain the reduced variables $\alpha, \vec{v}$, and Ψ semi-analytically, we can still check that the time-dependent physical variables $\rho, \vec{u}$, and Φ (or $\vec{B}$) when scaled properly by G, a, and t as in Eqs. (10-12), and plotted in the similarity coordinates ξ and θ, are invariant functions of time t. This exercise has been successfully performed and gives us confidence that the numerical codes used are accurate (see Allen, Shu, & Li 2003 for details).

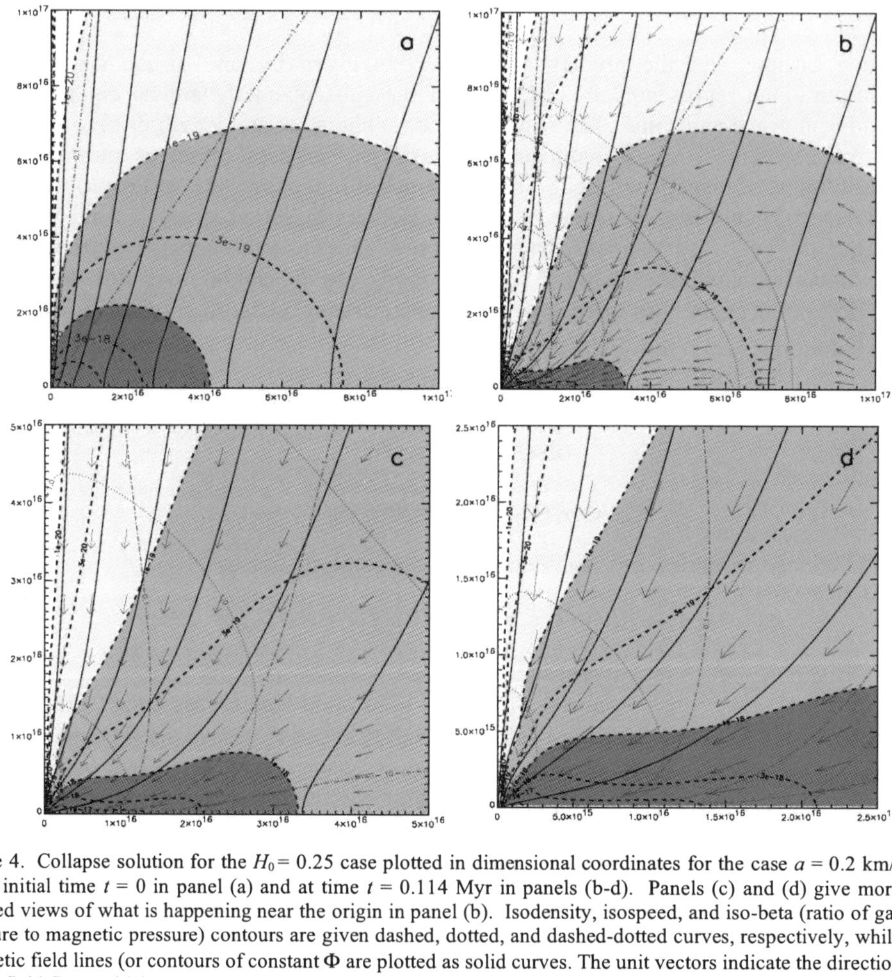

Figure 4. Collapse solution for the H_0= 0.25 case plotted in dimensional coordinates for the case a = 0.2 km/s at the initial time t = 0 in panel (a) and at time t = 0.114 Myr in panels (b-d). Panels (c) and (d) give more detailed views of what is happening near the origin in panel (b). Isodensity, isospeed, and iso-beta (ratio of gas pressure to magnetic pressure) contours are given dashed, dotted, and dashed-dotted curves, respectively, while magnetic field lines (or contours of constant Φ are plotted as solid curves. The unit vectors indicate the direction of the fluid flow, which has a component across field lines in the outer parts as self-gravity drags the material toward the center and is predominantly along field lines in the inner parts as the strong central concentration of field forces matter mostly to flow along field lines. (From Allen, Shu, & Li 2003.)

3.1. *Catastrophic Magnetic Breaking if Field Freezing Applies*

The effects of rotation can be included in a straightforward fashion, and the problem retains self-similarity if the rotation curve is a flat one,

$$\Omega r \sin\theta = \text{ constant } \equiv V_0, \tag{15}$$

at the pivotal instant $t = 0$ (because then one has only introduced another velocity scale into the problem). Surprisingly, when Allen, Li, & Shu (2003) carried out such simulations with realistic values of V_0 (a small fraction of a), they found that the results do not look very different from the case when there is no rotation ($V_0 = 0$). The answer is not as simple as that cases where $V_0^2 << a^2$ resemble the case $V_0 = 0$. While the inequality between rotational inertia and pressure is true for the outer part of the flow, it becomes irrelevant at small radii because (isothermal) gravitational collapse usually enhances the effects of rotation and diminishes the relative role of pressure.

In order of magnitude, we expect the effects of rotation to become important when the centripetal acceleration $u_\varphi^{\ 2} / r\sin\theta$ becomes comparable to the gravitational acceleration GM_* / r^2, where M_* is the mass of the protostar accumulated at the center. If we could ignore the effects of the magnetic field, spin angular-momentum per unit mass would be conserved in the non-viscous flow by Kelvin's circulation theorem. The combination $u_\varphi r\sin\theta$ would then retain its initial value $V_0 r_0 \sin\theta_0$ at the instant $t = 0$; whereas the mass accumulated at the center of the configuration is given approximately through Eq. (15) as $M_* \approx a^3 t / G$. Thus, for infall in the equatorial plane, $\theta = \theta_0 = \pi / 2$, we expect the centripetal acceleration to be balanced by gravity when $V_0^{\ 2} r_0^{\ 2} / r^3 \approx a^3 t / r^2$. The head of the expansion wave that initiates the infall races into the cloud at the speed of sound a and reaches a radius in time t given by at, whereas the equatorial material with the largest amount of specific angular momentum that falls into the central regions had an original location $r_0 \approx at / 2$. Thus, in the absence of viscous spreading, we anticipate the formation at time t of a centrifugally supported disk of reduced radius

$$\xi \equiv \frac{r}{at} \approx \frac{1}{4}\left(\frac{V_0}{a}\right)^2. \tag{16}$$

The simulations of Allen, Li, & Shu (2003) had enough dynamic range to have observed centrifugal disks of such sizes, yet none were found. Evidently, the twisting and stretching of the magnetic field as matter spirals toward the center is able to exert enough magnetic torque, through the propagation of torsional Alfven waves that carry away the spin angular momentum, to prevent the formation of a centrifugally supported disk. Figure 5 shows in 3-D the magnetic field configuration of a typical calculation. Notice that the field lines in the central regions, which resemble streamlines, spiral directly into the center rather than form a highly flattened, tightly coiled, structure that we might have expected with the formation of a centrifugally supported disk.

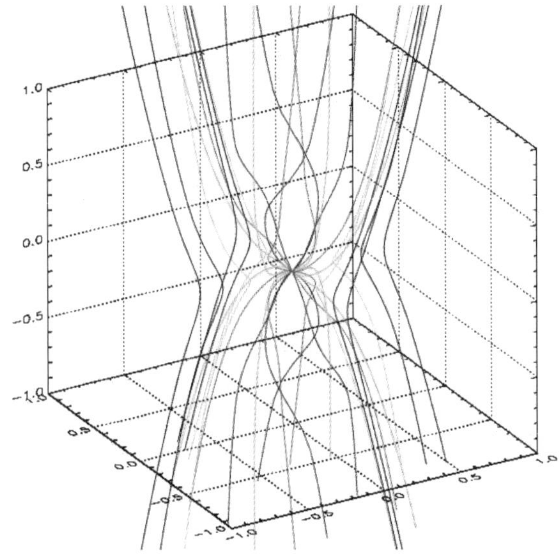

Fig.5. The 3-D magnetic configuration attained by a magnetized, rotating, and gravitationally collapsing, toroid plotted in similarity coordinates for the case H_0 = and $V_0/a = 0.25$. (From Allen, Li, & Shu 2003.)

3.2. *Field Freezing Produces a Split Monopole at the Center*

It is easy to see physically why the magnetic braking is so catastrophically efficient as to prevent the formation of centrifugal disks altogether if field freezing applies. When the magnetic flux that accompanies the mass is dragged into the protostar, the mass forms a monopole for the gravitational field and the flux forms a *split monopole* for the magnetic field. A split monopole differs from a monopole in that the field in one hemisphere points away from the star while in the other hemisphere it points toward the star. Because magnetic field lines never end (since $\nabla \cdot \vec{B} = 0$), each field line that enters the protostar in one hemisphere must leave it in the other. Outside of the origin, a pure monopole is a true vacuum field, but a split monopole has a current sheet that supports the oppositely directed field lines just above and below the equatorial plane. Because the field increases in strength toward the center as $B \propto 1/r^2$, and the density only increase as $\rho \propto r^{-3/2}$, the Alfven velocity $\propto B/\rho^{1/2} \propto r^{-5/4}$ increases relative to the free-fall velocity $\propto r^{-1/2}$ as the center is approached, $r \to 0$. If field freezing applies, the flow becomes sub-Alfvenic at small radii, which implies (1) the velocity vector must increasingly be channeled along the magnetic field, and (2) a strong lever arm exists in the split monopole to exert magnetic torque on the inflowing matter if this material contains a component of velocity in the azimuthal direction in addition to radial infall.

In similarity coordinates, it is possible to obtain analytically the inner solution for the problem in the limit of small ξ (Galli et al. 2006). We begin the discussion ignoring at first the presence of rotation. For $\xi << 1$, the reduced velocity corresponding to free-fall has the quasi-steady form

$$\mathrm{v}_{\xi} = -\left[\frac{2m_0(1+H_0)}{\xi}\right]^{1/2}. \tag{17}$$

The angular distribution for the flux function near the origin is

$$\phi = \frac{m_0(1+H_0)}{\lambda_*}(1-\cos\theta), \tag{18}$$

where λ_* is the dimensionless mass-to-flux ratio associated with the trapped flux in the protostar. Through numerical integration the quantity λ_* is uniquely defined as a function of H_0 if we assume ideal MHD, and it typically has a value of 1-4 for the range of H_0 of practical interest. Corresponding to Eq. (18) is a dimensional (radial) magnetic field (in the positive hemisphere) of form,

$$B_r = \frac{m_0(1+H_0)a^3t}{\lambda_* G^{1/2}r^2}, \tag{19}$$

with a reversal of sign across the equatorial plane appropriate for a split monopole. Field freezing in the present circumstances implies that the mass flux is proportional to the magnetic field, which yields for the reduced density at small ξ,

$$\alpha = \left[\frac{1}{2}m_0(1+H_0)\right]^{1/2}\xi^{-3/2}\beta(\phi), \tag{20}$$

where $\beta(\phi)$ is a function describing how matter is loaded on field lines and can be obtained from the inner limit of an outer solution obtained by numerical simulation for each case of H_0. Figure 6 shows the infall solution at small ξ in similarity coordinates obtained from a procedure of matched asymptotic expansions for a range of values for H_0.

When one includes rotation into the problem, $V_0 \neq 0$ in the pivotal state, the patterns of magnetic field, flow velocity, and density configuration in the meridional plane remain little changed from the non-rotating case $V_0 = 0$. The reduced azimuthal velocity and magnetic field can be obtained analytically from the other variables through the constraints of field freezing and angular momentum transport. The latter is given by a conservation relation that expresses the following fact. In quasi-steady state (as applies in the inner regions which are crossed by fluid flow in a time short compared to the evolutionary time of the problem), any angular momentum lost by the matter must be transferred outward by the magnetic torques in the problem. The equations then show that streamlines and field lines spiral into the protostar at the center, with the number of azimuthal turns in the spiral being typically small for realistic choices of the parameters H_0 and V_0/a (see the discussion in Galli et al. 2006). This calculation then provides the proof that the formation of centrifugally supported disks is suppressed if field freezing were to apply in the magnetized, rotating, gravitational collapse of realistic molecular cloud cores.

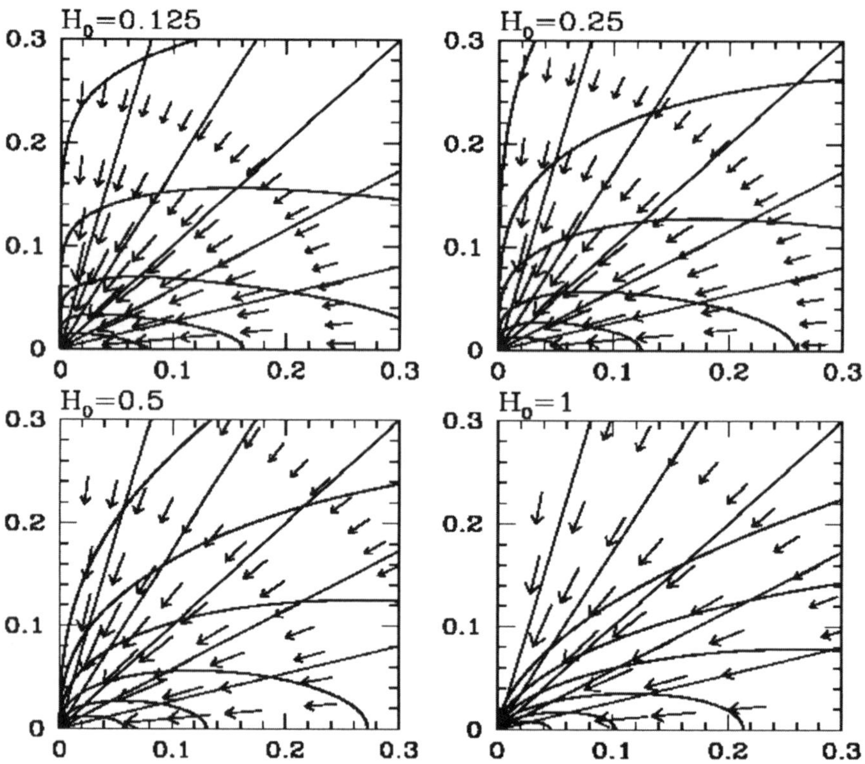

Fig.6. Isodensity contours (solid curves), field lines (fan of straight lines from origin), and unit velocity vectors (arrows) plotted in similarity coordinates for small ξ in the inner regions of gravitational infall for a variety of H_0 cases with zero rotation (From Galli et al. 2006.)

There exists another empirical reason why field freezing cannot apply to the real problem. The dimensional mass-to-flux ratio in the central star is given by the expression

$$\frac{M_*}{\Phi_*} \equiv \frac{\lambda_*}{2\pi G^{1/2}}. \tag{21}$$

If we know the dimensionless ratio λ_*, then we can calculate the flux that threads one hemisphere of the star if we are given its mass M_*. If λ_* were given by field freezing in gravitational collapse from the pivotal state, i.e., if λ_* had a value of 1-4, then the resulting surface fields, obtained by dividing Φ_* by the cross-sectional area of the star, are four orders of magnitude larger than measured observationally to exist in newly born T Tauri stars. The basic problem was known to Mestel and Spitzer (1956) and motivated them to propose that flux loss through ambipolar diffusion in the interstellar medium might be the resolution to the problem. The developments of more recent studies show that flux loss by ambipolar diffusion drives molecular cloud cores to an unstable pivotal state that leads to dynamical collapse to much smaller dimensions where the flux-to-mass ratios are still too large by four orders of magnitude to be consistent with observations without additional dissipation of the magnetic field at the smaller dimensions.

4 Effects of Non-ideal Magnetohydrodynamics

4.1. *Breakdown of Ideal MHD at Small Scales*

Consider the magnetic configuration of a collapsing cloud core in which field freezing applies. Topologically, we may partition the magnetic fields into two categories: those that have been dragged into the protostar with the infalling matter, and those which have not yet experienced this stretching. The latter field lines cross the midplane vertically and link the upper side with the lower. A separatrix in the shape of a sideways "Y" in the meridional plane divides the first category of field lines from the second. The inner solutions shown in Fig. 6 refer to the region enclosed schematically by the dotted circle in Fig. 7 where the magnetic configuration takes the form of a split monopole to zeroth approximation. We wish to begin our discussion of non-ideal MHD by focusing our attention on the current sheet represented by the sudden reversal in the direction of the magnetic field as one crosses the mid-plane. Our concern will be the destruction of the current sheet by the effects of finite resistivity and the resulting magnetic reconnection that occurs inside the dotted circle. The release by magnetic reconnection of the anchor point of the interior fields occurring near the center will affect the exterior configuration, perhaps via the propagation of shock waves traveling outwards in the equatorial plane. The back reaction can be likened to a stretched rubber band which breaks and reconnects, a problem that we leave for future investigations (cf. Li & McKee 1996).

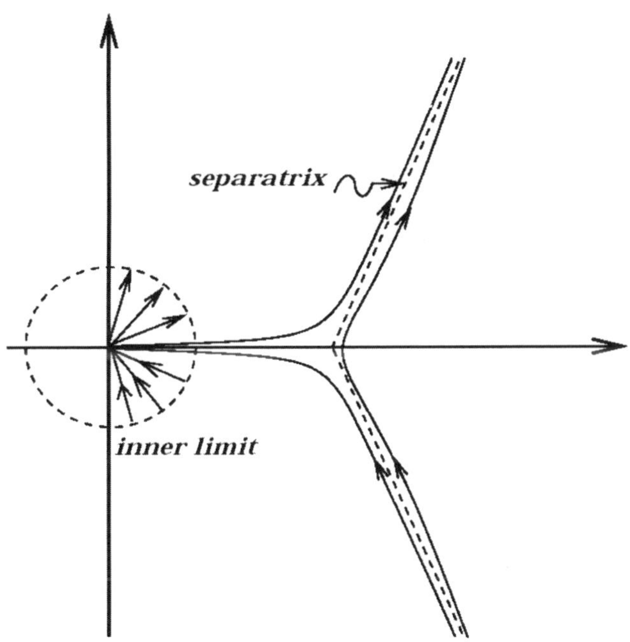

Fig. 7. Schematic drawing of the magnetic configuration that arises when the concepts of ideal MHD are applied to the gravitational collapse of magnetized molecular cloud cores. (From Galli et al. 2006.)

4.2. *Ohmic Dissipation Yields Central Region with Low Uniform Field*

Consider the generic equation for magnetic field diffusion,

$$\frac{\partial \vec{B}}{\partial t} + \nabla \times \left(\vec{B} \times \vec{u}\right) = -\nabla \times \left(\eta \nabla \times \vec{B}\right), \tag{22}$$

where η is nominally the coefficient of electric resistivity, but which can be generalized to include the effects of ambipolar diffusion for axisymmetric configurations:

$$\eta_{\text{eff}} \equiv \eta + \frac{B^2 \tau}{4\pi\rho}. \tag{23}$$

In what follows, we shall write η for notational simplicity and refer to this quantity as the "electric resistivity," but we reserve for detailed application the possibility that we really mean η_{eff} which contains a contribution from ambipolar diffusion in the form given by the square of the Alfven speed $B^2 / 4\pi\rho$ times the ion-neutral collision time τ.

To simplify the mathematical discussion, assume that η is a constant in equation (22) and that $\vec{u}$ is given at small radii by radial free-fall,

$$u_r = -\sqrt{\frac{2GM_*}{r}}, \tag{24}$$

where M_* is the mass of the star and increases on an accumulation time that is long compared to the dissipation time scale of interest here. In this case, we may look for quasi-steady solutions of equation (22) that involves a balance of the advective term on the left-hand side and the dissipative term on the right-hand side. In order of magnitude the diffusion time across a region of size r associated with diffusivity η is $t_{\text{diff}} = r^2 / \eta$. For this time to be comparable to the free-fall time across r at the speed given by Eq. (24), $t_{\text{ff}} = r / u_r$, the characteristic size must be given by the expression,

$$r_{\text{Ohm}} = \frac{\eta^2}{2GM_*}, \tag{25}$$

which we call the Ohmic radius. For the values $\eta = 2 \times 10^{20}$ cm^2 s^{-1} and $M_* = 1\, M_\odot$ (see below), we get $r_{\text{Ohm}} = 8.4$ AU and $t_{\text{diff}} = 3$ yr, which is much less than the evolutionary time of a few hundred thousand yr that characterizes collapse from the pivotal state. Thus, the empirical justification for the quasi-stationary analysis is a good one.**

**However, empirically, one also knows from observations of the Sun that the magnetic reconnection of a region with a "Y" type separatrix occurs not quasi-statically but in flares. This analogy alerts us to the possibility that the same instability, or a related one, might happen in protostars, and indeed X-ray flares and larger disturbances of longer duration called FU Orionis outbursts are a common occurrence in all newly-born sunlike stars.

Since the governing partial differential equation is linear in $\vec{B}$, we may solve it by the usual method of separation of variables in r and θ, subject to the condition that the field looks like a split dipole at "infinity" (really, the "outer limit of an inner solution") and is regular at the origin. The axisymmetric solution with no rotation can then be found as an infinite series expansion in Legendre polynomials in $\cos\theta$ and confluent hypergeometric equations in the variable r / r_{Ohm}, where r_{Ohm} is given by Eq. (25). Figure 8 depicts the solution in the meridional plane using r_{Ohm} as the unit of length.

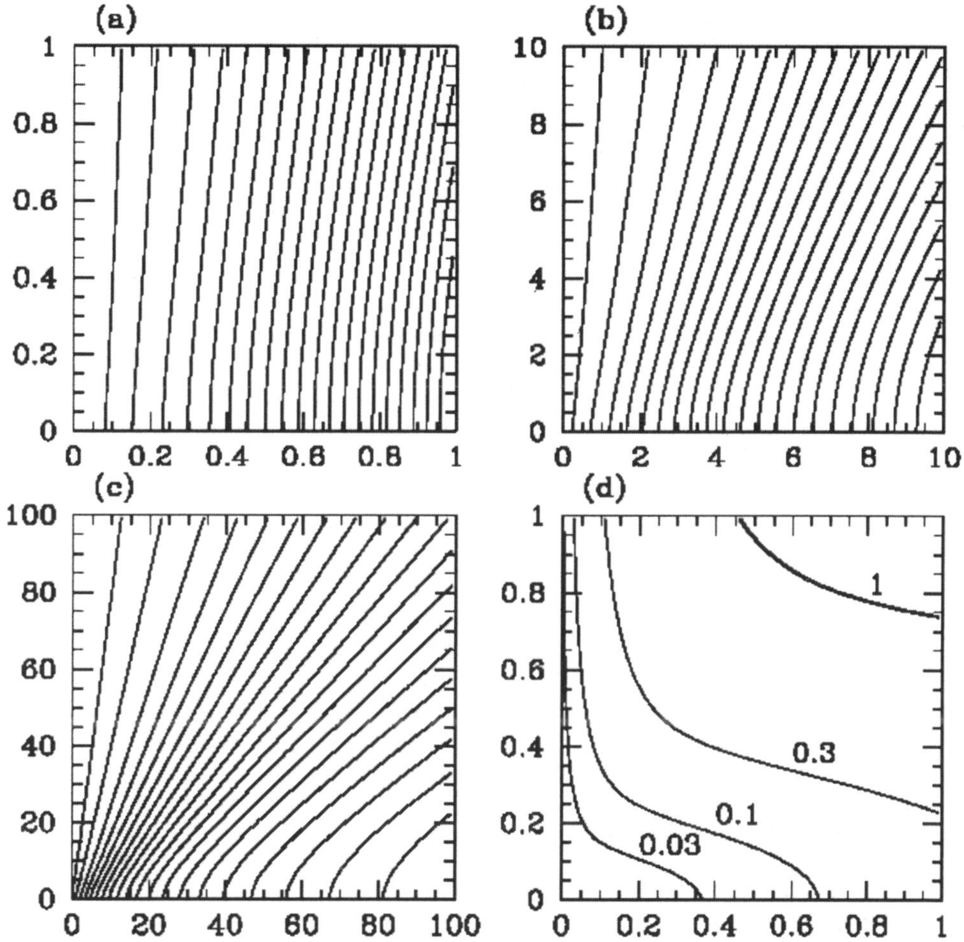

Fig. 8. Magnetic field lines inside equatorial and axial distances of (a) 1, (b) 10, and (c) 100 r_{Ohm}. In (d), we plot the ratio of the magnitudes of the Lorentz force and the gravitational force. Notice that in the equatorial region this ratio is small, leading to a self-consistent expectation that we can expect a centrifugally supported disk to form there when the effects of rotation are present. (From Shu, Galli, Lizano, & Cai 2006.)

Starting from a configuration that looks like a split monopole at scales large compared to $r_{\rm Ohm}$, the magnetic field becomes straight and uniform at scales small compared to $r_{\rm Ohm}$ (see Fig. 8). Analytically, the field in the innermost regions reaches a central value that is given by the expression,

$$B_{\rm c} = \frac{\Phi_*}{30\pi {r_{\rm Ohm}}^2} = \frac{4G^{5/2}{M_*}^3}{15\lambda_* \eta^5}, \tag{26}$$

where Φ_* and λ_* are the magnetic flux and mass-to-flux ratio, respectively, that would have obtained in the star *if field freezing had held.* For fiducial values of $\lambda_* = 2$, $\eta = 2\times10^{20}\,\rm cm^2\ s^{-1}$, and $M_* = 1\ M_\odot$, we get $B_{\rm c} = 1$ gauss, which is consistent with the remnant magnetization levels of primitive meteorites for the magnetic fields resident in the solar nebula at the time of the formation of the Sun and planetary system. Indeed, it is the constraint provided by chondritic meteorites that led us in the first place to choose $\eta = 2\times10^{20}\,\rm cm^2\ s^{-1}$ as a characteristic value for the electrical resistivity associated with the formation of the solar nebula (a centrifugally-supported accretion disk).

In fact, the adopted η exceeds the highest values theoretically computed for microscopic collisional values under conditions believed to apply to the primitive solar nebula. In making this statement, we recognize that the theoretical values are highly uncertain as they depend sensitively on the assumptions made about the clumping of interstellar grains that have fallen into the solar accretion disk. The processes and time scales by which dust grains aggregate to become planetesimals and planets are highly controversial. Nevertheless, we are impressed with the fact that the requisite (mean) value of the electric resistivity needed realistically to break the assumption of field freezing and avoid the catastrophic magnetic braking that would otherwise prevail is larger than the largest values quoted in the theoretical literature by at least one order of magnitude (see the discussion in Shu et al. 2006). Once an accretion disk has formed and established itself, the lower theoretical values of η are acceptable because radial inflow through an accretion disk is much slower than free fall. The requisite diffusion time for the magnetic field can be correspondingly longer in the same circumstances for a given characteristic scale r. Turbulent diffusion and/or large sporadic eruptions may be required to mediate the competing demands for a relatively rapid dissipation of advected magnetic field in gravitational infall and the relatively slow dissipation that may be available in fully formed YSO accretion disks.

5 X-winds and Protostellar Jets

We come now to the last topic of this review paper: the jets and bipolar outflows that are observed in all YSOs at an early enough stage in regions of low-mass star formation. It is now almost universally accepted that jets and bipolar outflows are driven by processes that intimately involve a combination of rapid rotation and magnetic fields. A centrifugally supported disk is an ideal object to supply the rapid rotation because, except for small thermal and magnetic stresses, each radius is rotating, by definition, at the fastest rate consistent with mechanical equilibrium. The source of the embedded magnetic field is more controversial. One school of thought postulates that it is associated with the disk itself, brought in from the interstellar medium by the process of gravitational collapse and accretion described in the previous sections (see, e.g., the review of Konigl & Pudritz 2000). Another school of thought holds that while the rapid rotation is associated almost certainly with a (magnetically viscous) accretion disk, the strong magnetic field is generated by dynamo action in the rotating, vigorously convecting, stars that characterize the protostellar and pre-main-sequence stages of evolution of young sunlike stars. The central motivation for the latter point of view is the observation that without such strong stellar fields, we cannot begin to understand how the viscous in-spiral of matter from a highly flattened accretion *disk* can produce a relatively slowly-rotating normal star that is a *sphere* (see the review of Shu et al. 2000).

5.1. *Interaction of an Accretion Disk and a Strongly Magnetized Central Star*

Figure 9 shows the magnetic field lines that result in steady state from the idealized MHD interaction of an infinitesimally thin, conducting, accretion disk with the magnetosphere of a young star that is approximated as being a point dipole at the origin if the disk had not been there. Because the sound speed is small compared to the resulting flow speeds, the effects of thermal pressure have been ignored in the asymptotic analysis (see Shu et al. 1994). The unit of length used in the diagram is the inner edge of the disk,

$$R_x = \phi_{dx}^{-4/7}\left(\frac{\mu_*^4}{GM_*\dot{M}_d^2}\right)^{1/7}, \tag{27}$$

where μ_* is the unperturbed magnetic-dipole moment of the star, $\dot{M}_d$ is the mass accretion rate through the disk, and ϕ_{dx} is a dimensionless number of order unity that characterizes how much magnetic flux of the unperturbed dipole has been trapped in the fan of field lines seen at the edge of the accretion disk in Fig. 9 at dimensionless radius equal to 1. When values characteristic of T Tauri stars are put into Eq. (27), R_x turns out typically to be about 5 times the stellar radius R_* (not depicted in the diagram).

It can be shown that poloidal magnetic fields of sufficient strength threading a cold disk at an angle with respect to the horizontal less than 60° provide enough centrifugal fling to drive an MHD flow out of the equatorial plane of the disk (Chan & Henriksen 1980, Blandford & Payne 1982). In the inward and outward directions, these field lines, depicted in a different shade in Fig. 9, provide "guide wires" for a funnel flow (Ostriker & Shu 1995) onto the star and an X-wind (Najita & Shu 1994) away from the star that emanates from the X-point R_x at the inner edge of the cold disk. Sandwiched in between wind and funnel-flow are dead-zone field lines (heavy black curves) that thread the disk too nearly vertically to have sufficient fling in either the outward or inward directions. Thus, in the 180° range of angle in the two half-planes above and below the disk, each sector – wind, dead zone, and funnel flow – occupies an elegant 60° in the approximation of a cold MHD flow.

The gas that falls onto the star has angular momentum removed from it by the funnel flow field lines so that by the time it joins the star, it is rotating at the slow average rate of the (nearly spherical) star. The excess angular momentum, which characterized the original orbital angular momentum of the inner edge of the disk, is transferred to the

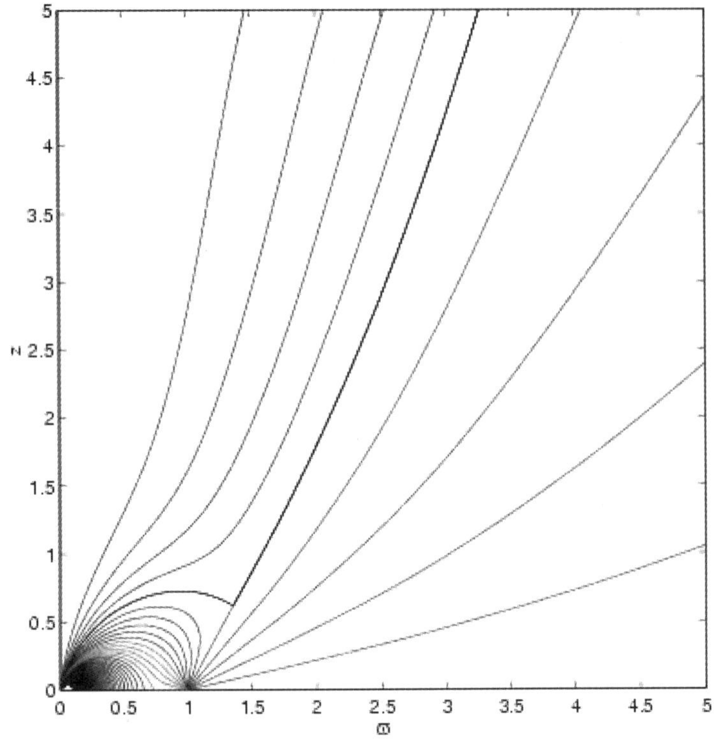

Figure 9. Configuration of magnetic field achieved in an axisymmetric, time-independent, calculation of the interaction of a stellar magnetosphere with a cold accretion disk, approximated to be infinitesimally thin and lying along the dimensionless ϖ axis beyond $\varpi = 1$. The X-wind flow occurs along the lightly shaded field lines that emanate from the outer third of a fan of field lines at the X-point ($\varpi = 1, z = 0$) and head off to the right of the diagram. The funnel flow occurs along the somewhat darker shade of field lines that emanate from an inner third of the fan of field lines at the X-point and land on the surface of the star, which is small sphere centered on the origin that can be drawn in after the fact for the assumption of a pure unperturbed stellar dipole. The field lines in the dead zone (light shade of gray) between the wind and funnel flow, rise too vertically from the plane of the disk to be dynamically active. Field lines that appear to enter the star from above are the dead counterparts of the original dipole field lines that have been opened by the X-wind. Since magnetic field lines never begin or end, they are equal in number to the field lines in the X-wind and connect with them, so to speak, at "infinity" in a steady calculation. (From Shu et al. 2000.)

material remaining in the disk, and this transfer tries to drive the inner edge outward, which truncates the inner edge to a radius R_x larger than R_*.

An opposite action/reaction occurs in the X-wind. On the latter field lines, as the gas flows upward and outward, it is forced to rotate faster than the local speed for centrifugal balance, which accelerates it into an outflowing wind. The angular momentum given to the X-wind gas is supplied by the remaining material at the footpoint of the corresponding field lines in the disk, which drives such material inward at a faster rate than given by the action of (turbulent) viscous accretion alone. In steady state, the inward drift of the inner edge of the disk promoted by the back reaction to the X-wind flow is balanced by the outward drift induced by the disk's reaction to the funnel flow and produces an equilibrium value for R_x given by Eq. (28). The resultant pinch that squeezes inward and outward on the field lines in the X-region provides the fan at the X-point that began our discussion of the simplest version of X-wind theory.

The conditions of mass and angular momentum balance of the matter that passes through the X-point provides one more constraint, which governs the complementary fractions f and $1-f$ of the disk accretion rate $\dot{M}_{\rm d}$ that end up as wind and funnel flow onto the star: $\dot{M}_{\rm w} = f\dot{M}_{\rm d}$, $\dot{M}_* = (1-f)\dot{M}_{\rm d}$. In terms of the angular momentum per unit mass carried by the wind matter to infinity, and by the funnel-flow matter to the star, as measured in units of the disk orbital value at $R_{\rm x}$, the fraction f is given by the expression:

$$f = \frac{1-\bar{J}_* - \tau_{\rm d}}{\bar{J}_{\rm w} - \bar{J}_*}, \tag{28}$$

where $\tau_{\rm d}$ is the angular momentum per unit mass in the same units transferred outward from the inner edge of the disk through viscous torques (Shu et al., 1988, 1994). Mechanistically, the wind fraction f is determined by complex non-ideal MHD effects that apply at the lightly ionized disk edge (see Section 1.3). Without investigating the details of the process by which matter accreting through the disk becomes loaded onto wind or funnel-flow field lines, we may construct an approximate physical argument to guess the wind fraction f. The trick is to recognize that the diffusive processes in the model occur within a narrow range of radii (infinitesimal in lowest asymptotic order). Thus, conditions, however complicated they may be, are the same for the loading of the wind, dead zone, and funnel flow, each of which contain an equal number of field lines since the totality of trapped field lines in the X-region is distributed uniformly with respect to the local polar angle. Lacking other information, therefore, the simplest guess is that at any instant of time in a quasi-steady description, only 1/3 of the accretion flow manages to climb onto wind field lines, the other 2/3 pass onto either dead zone or funnel flow field lines, eventually to end up on the star. This naïve guess produces $f = 1/3$.

To get this answer from Eq. (28), we suppose that $\bar{J}_*$ is much less than unity (otherwise, the star would not end up quasi-spherical), and that $\tau_{\rm d}$ is also $\ll 1$ (otherwise, it would be hard to understand disk truncation). Equation (28) now yields the estimate $\bar{J}_{\rm w} \cong 3$. An astronomical fact is available to check this deduction. Energy conservation (or Bernoulli's theorem) associated with the system of partial differential equations shows that the mean dimensional terminal speed reached on a typical streamline or field line is given for a cold flow by the equation:

$$u_{\rm w} = \left(2\bar{J}_{\rm w} - 3\right)^{1/2} \Omega_{\rm x} R_{\rm x}, \tag{29}$$

where $\Omega_{\rm x} R_{\rm x}$ is the Keplerian rotation speed at the inner edge of the disk. The formula given above yields a couple hundred km/s as a typical estimate for the terminal velocity of X-winds, consistent with the available observational data for observed YSO jets. The quasi-stationary hypothesis also requires that the coefficient $\phi_{\rm dx}$ in eq. (27) be computed by assuming rotation periods for the central stars that magnetically lock them in synchronism with the inner edge of the disk. The implied stellar periods lie within a range of several days, again in agreement with the available observational evidence.

Before leaving this topic, we note that the physical discussion of this subsection has not used the *details* of the solution of the partial differential equation that governs the spatial distribution of magnetic field lines (and streamlines) in the meridional plane (i.e., the result plotted in Fig. 9), which is an elegant mathematical problem in its own right, but one which we have no space to discuss here. We merely note that the *idea* of asymptotics is by itself such a powerful tool that it can supply general principles to steer us through complex and tricky physical issues that might otherwise bog down an analysis whose main purpose, as Prof. Lin emphasized at this meeting, is to put the stress on "applied" and not on "mathematics."

5.2. *Asymptotic Behavior of X-wind at Large Distances from the Source*

Figure 10 shows isodensity contours (solid curves) and streamlines (dished curves coinciding with field lines in the meridional plane) of a typical case of the X-wind on scales of 10, 100, 1,000, and 10,000 times R_x. The magnetic fields wrap up in the azimuthal direction out of the meridional plane because of the difference in angular rates

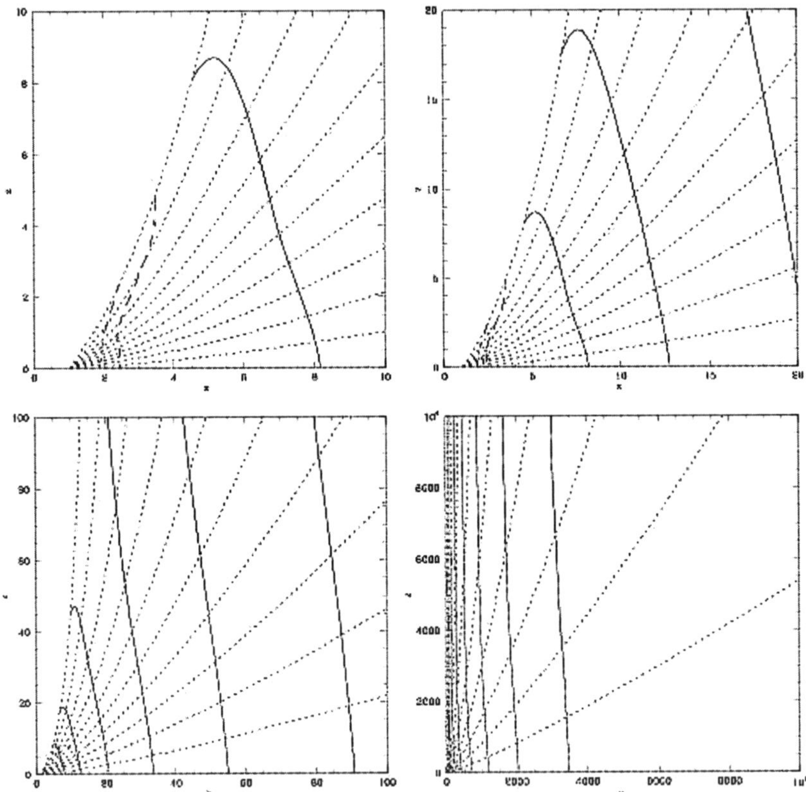

Figure 10. Streamlines (dished curves) and isodensity contours (solid curves) in a typical solution for the X-wind at a variety of spatial scales. The empty space along the vertical axis is a "hollow cone" filled with vacuum field lines drawn out from the star's original dipole distribution (see Fig, 9). The X-point on the horizontal axis that begins the flow is formally where a smooth crossing of the slow MHD speed (zero to the lowest order of approximation for vanishing sound speed) occurs; the dashed lines indicate the surfaces for smooth Alfven and fast MHD corssing. The interface between the longitudinal field of the hollow cone and the mostly toroidal field of the uppermost streamline acquires a shape that becomes asymptotically parabolic in the meridional plane because of (magnetic) pressure equilibrium across the interface. (From Shu et al. 1995.)

of rotation of the wind and the inner edge of the disk. The associated "hoop stresses" lead to a logarithmically slow collimation of streamlines toward the rotation (z) axis (Shu et al. 1995). Thus, although about 50% of the streamlines are heading within 6° of the rotation axis at the largest scale shown in Fig. 10, another 50% are still directed toward shallower angles that include the equatorial direction. The effect is a consequence of Newton's third law: whatever (magnetic) action there exists to force some streamlines up toward the rotation axis, there must be an associated back reaction that force other streamlines down toward the equatorial plane . These facts mean that the streamlines *per se* do not appear very jetlike until the flow gets to a few orders of magnitude larger scale yet. Observationally, however, stellar jets look very rectilinear on scales that can be resolved by the Hubble Space Telescope (see second panel of Fig. 11), which, for the nearest star-forming regions, lie within the third panel of Fig. 10 that has 1,000 as its largest dimension. Why is there this apparent conflict between theory and observations?

5.3. *Optical Illusion of Jet when Outflow is Lit by Turbulent Shockwaves*

The answer to the last question can be found, we believe, in the isodensity contours of Fig. 10. For incompletely understood reasons, the asymptotic solutions of X-wind (and to a lesser extent, disk wind) theory have nonlinear attractors that produce isodensity contours which become cylindrically stratified at large distances independent of the details of what is happening near the launch region (see, e.g., Li 1996). In the specific case of X-wind theory, ρ approaches $\rho \propto \varpi^{-2}$ independent of z for large ϖ.

Now, astronomers look at radiation, and not at streamlines. Astrophysical gases that are heated sufficiently produce emission lines that are sensitive to the density and temperature of the medium. If the gases are thermostated by radiative cooling to more or less uniform temperatures, the iso-brightness contours (called "isophotal contours" in astronomical jargon) will resemble isodensity contours more than they do streamlines. Observationally, even if a gas is not heated to uniform temperatures (e.g., by shockwaves propagating in the medium), the subsequent radiative cooling behind the shock fronts tend to select out for any particular radiation diagnostic a fairly narrow range of temperatures.

The left-hand panel of Fig. 11 shows a synthetic image of the X-wind model of Fig. 10 heated by a volumetric energy source with the functional dependence

$$\Gamma = \alpha \frac{\rho u^3}{s}, \tag{30}$$

where α is a dimensionless free-parameter of the semi-empirical approach and u is the flow speed of the X-wind at a distance s along the streamline measured from the launch point at ($\varpi = R_x, z = 0$). Equation (30) is a heuristic formula that suggests the presence of propagating internal shockwaves or hydrodynamic turbulence in the problem, perhaps both. The fact that good radiative intensities are obtained for the relative cooling lines when α in the heuristic formula is chosen to have a very small value, 0.002 to 0.01, suggests that any shockwaves or turbulence present in these hypersonic flows are relatively weak. This conclusion justifies our neglect of the back-reaction on the mean flow.

The isophotal contours in the synthetic image are comparable to the levels seen in the knots of the observed astronomical jet in the right panel of Fig. 11, but the color coding (changed to a grayscale in the published figure) is different. The observed jet is probably pulsing and perhaps turbulent. Nevertheless, the trends of brightness and kinematics (not shown) are adequately modeled by the (laminar) X-wind theory augmented by the *ad hoc* heating function of Eq. (30) in the energy equation of radiation magnetohydrodynamics.

When one looks at the optical image of the astronomical observations alone, as is frequently the case in textbook presentations, one is tempted to interpret the picture in terms of a pencil beam of gas particles flowing rectilinearly in the *z*-direction; hence the association of the name "jet" with the phenomenon. X-wind theory, which produces a less well-collimated assortment of streamlines, regards the effect at the moderate length scales of many observations as an "optical illusion" created by cylindrically stratified isodensity contours.

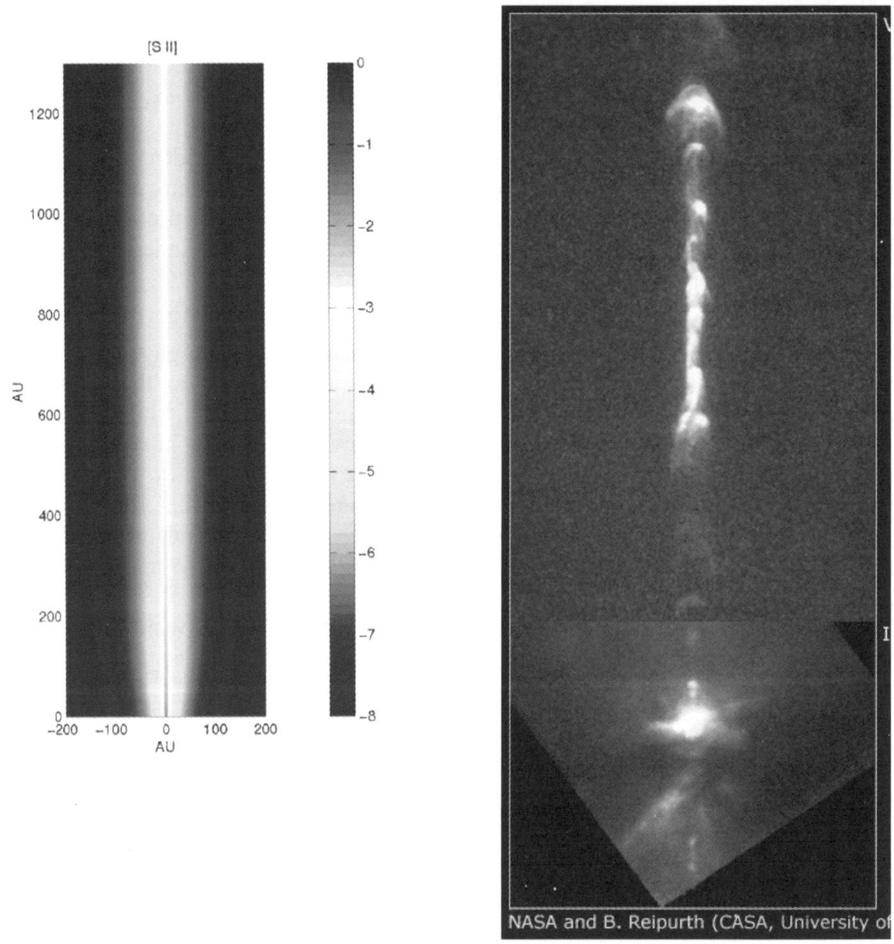

Figure 11. Synthetic image of an X-wind (from Shang et al. 2002) and an astronomical image of a jet from a young stellar object (photo credit: NASA and Bo Reipurth, University of Colorado).

6 Summary

The large dynamic range inherent in many problems of star formation imply that semi-analytic attacks based on asymptotic methods can still be competitive with numerical simulations despite the many orders of improvement in computational speed and storage capacity of electronic computers in the past few decades. Time-dependent calculations in three spatial dimensions are probably best left to numerical simulations, but for problems of lower dimensionality where physical insight and not numbers is the goal, applied mathematics of the variety taught and practiced by C. C. Lin still has an important role to play in the modern physical and biological sciences. For the transport problems in the non-ideal magnetohydrodynamics of star formation discussed in this lecture (and not discussed, such as "viscous" disk accretion), turbulence raises its frightful head at each juncture. Further progress in the field depends on taming this beast. Here, I take heart in the lesson that the early career of C. C. Lin taught, which is that some aspects of the problem of turbulence are also amenable to attack by asymptotic methods.

Acknowledgments

My work in the field of star formation was supported by the National Science Foundation and the National Aeronautical and Space Administration in the United States and by the National Science Council and the Foundation for Outstanding Scholarship in Taiwan. Inspiration and encouragement came from a large numbers of coworkers, postdocs, and students, but most notably from my teacher and mentor, C. C. Lin, to whom I dedicate this paper.

References

Allen, A., Li, Z.-Y., & Shu, F. H. *Ap J,* **599**, 363 (2003).
Allen, A., Shu, F. H., & Li, Z. Y., *Ap J*, **599**, 351 (2003).
Blandford, R. D., & Payne, D. G., *Mon Not Roy Astr Soc*, **199**, 883 (1982).
Chan, K. L., & Henriksen, R. N. , *Ap J*, **241**, 145 (1980).
Desch, S. J., & Mouschovias, T. Ch., *Ap J,* **550,** 314 (2001).
Fatuzzo, M., & Adams, F. C., *Ap J*, **570**, 210 (2002).
Galli, D., Lizano, S., Shu, F. H., & Allen, A., *Ap J,* in press (2006).
Jijina, J., Myers, P. C., & Adams, F. C., *ApJ Supp*, **125**, 161 (1999).
Konigl, A., & Pudritz, R. E., in *Protostars and Planets IV*, eds Manning, V., Boss, A. P., & Russell, S. S. Tucson, Univ. of Arizona Press, p. 759 (2000).
Li, Z.-Y., *Ap J*, **473**, 873 (1996).
Li, Z.-Y., & McKee, C. F. , *Ap J*, **464,** 373 (1996).
Li, Z.-Y., & Nakamura, F., *Ap J*, **609**, 83 (2004).
Li, Z.-Y., & Shu, F. H., *Ap J*, 468, 261 (1996).
Lizano, S., & Shu, *ApJ*, **342**, 834 (1989).
Mestel, L., & Spitzer, L.1956, *Mon Not Roy Astr Soc*, **116**, 503 (1956),
Myers, P. C., & Benson, P. J., *Ap J*, **266**, 309 (1983).
Myers, P. C., & Lazarian, A., *Ap J Lett*, **507**, 157 (1998).
Najita, J. R., & Shu, F. H., *Ap J*, **429**, 808 (1994).

Nakano, T., *Pub Astr Soc Japan*, **31,** 697 (1979).
Ostriker, E. C., & Shu, F. H., *Ap J*, **447**, 813 (1995).
Shang, H.., Glassgold, A., Shu, F. H ., & Lizano, S., *Ap J*, **564**, 853 (2002).
Shu, F. H., *Ap J*, **214**, 798 (1977).
Shu, F. H., Adams, F. C., & Lizano, S. *Ann Rev Astr Ap*, **25**, 23 (1987).
Shu, F. H., Galli, D., Lizano, S., & Cai, M. J., *Ap J*, in press (2006).
Shu, F. H., Lizano, S., Ruden, S. P., & Najita, J., *Ap J Lett*, **328**, 19 (1998).
Shu, F. H., Najita, J., Ostriker, E. C., & Shang, H., *Ap J Lett*, **455**, 155 (1995).
Shu, F. H., Najita, J., Ostriker, E., Wilkin, F., Ruden, S., & Lizano, S., *Ap J,* **429**, 781 (1994).
Shu, F. H., Najita, J. R., Shang, H., & Li, Z.-Y., in *Protostars and Planets IV*, eds Manning, V., Boss, A. P., & Russell, S. S., Tucson, Univ. of Arizona Press, p. 789 (2000).
Zweibel, E., *Ap J,* **567,** 962 (2002).

SOLITARY WAVES FROM OPTICS TO FLUID DYNAMICS

M. J. ABLOWITZ, A. DOCHERTY

Department of Applied Mathematics, University of Colorado at Boulder, Boulder, CO 80309-0526, USA

Localized wave solutions, such as solitary waves, often termed solitons, arise naturally in nonlinear optics and fluid dynamics. In optical communications, asymptotic analysis leads to the "classical" and "dispersion-managed" (DM) nonlocal nonlinear Schrödinger equations which have localized pulses as special solutions. Recent research has shown that similar dispersion-managed equations occur in mode-locked lasers. A numerical method is introduced to find these and other localized waves in nonlinear optics, water waves and multi-fluid systems. In water waves the classical equations are reformulated as a system of two equations, one of which is an explicit nonlocal equation, for the wave height and for the velocity potential evaluated on the free surface. The nonlocal system is shown to reduce to known asymptotic limits in shallow and deep water. Included in these asymptotic reductions are the Boussinesq, Benney-Luke and nonlinear Schrödinger equations. Two dimensional lumps with sufficient surface tension, are obtained numerically.

1. Introduction

The studies of nonlinear optics and water waves are wide ranging disciplines with broad interest; the literature is extensive and in addition to research papers and memoirs there are many books which deal with this subject, [cf 1–4]. In many cases, finding localized waves to the governing equations are of keen scientific interest. Perhaps the best known example goes back to J. Scott Russell [5] with his observation of solitary waves in shallow canals in Edinborough, Scotland. This provided motivation for the seminal works of Boussinesq [6] and of Korteweg-deVries (KdV) [7] in shallow water, who derived approximate equations governing water waves which possess special solitary wave solutions.

In nonlinear optics solitary waves, usually called solitons, also play an important role. In optical fibers solitons were predicted to exist in 1973 [8, 9]. They were subsequently demonstrated experimentally by Mollenauer et al. [10]; [see also 11, 12, for additional references and historical background]. In order to improve and enhance optical communications various technological difficulties had to be overcome; some of the more important developments include: i) all-optical amplifiers which are used to counteract damping; and ii) dispersion-management (DM) [13, 14] in which strong variations in the underlying dispersion of the fibers are employed to significantly reduces penalties (e.g. reducing the effects of four wave mixing [15, 16] and reducing the size of frequency and timing shifts [17]) that arise in multi-channel communications systems. DM systems have been of keen technological interest and have been recently installed in commercial communications environments. Dispersion-managed systems also give rise to localized, i.e. soliton, solutions [14]. In Ablowitz and Biondini [18] a nonlocal equation which governs strongly dispersion-managed communications systems was derived by the asymptotic method of multiple scales. This asymptotic equation, termed the DMNLS (dispersion managed nonlinear Schrödinger equation) equation admits localized solutions which were found numerically in Ablowitz and Biondini [18], [see also 19, 20], and were rigorously proven to exist in Zharnitsky et al. [21]. Interestingly DM systems also give rise to other special solutions of the DMNLS equation, in particular quasi-linear pulse solutions [22, 23].

The numerical method, employed in Ablowitz et al. [18, 19], to investigate soliton solutions of the DMNLS equation, is based upon taking the Fourier transform of the nonlinear equation, introducing a convergence factor, and then iterating the resulting equation until convergence to a fixed point in function space. The method was first introduced in 1976 by Petviashvili [24]. The convergence factor depends on the homogeneity of the nonlinear terms. The technique works well for problems with a single polynomial nonlinear term [see also 25]. However, although many interesting systems have a single nonlinear term, this is clearly a special situation. Recently, another way to find localized waves was introduced [26]. The main ideas are to go to Fourier space (this part is the same as Petviashvili [24]), then renormalize variables and obtain an algebraic system coupled to the nonlinear integral equation. We have found the method of coupling to be remarkably effective and straight forward to implement. The localized mode is determined from a convergent fixed point iteration scheme.

The numerical technique, called spectral renormalization (SPRZ), is reviewed in this paper and employed in order to find localized waves to a variety of nonlinear problems that arise in nonlinear optics and fluid dynamics. In optics we revisit the nonlocal DMNLS equation where the DM soliton solutions were first obtained. Next we discuss another nonlocal system of equations which arises in both optics and water waves. The system is termed: as NLSM (nonlinear Schrödinger equation with mean) equation. In nonlinear optics it arises in quadratic nonlinear media under the quasi-monochromatic assumption. In [27, 28] it is shown that now with quadratic nonlinearity introduced, the classical nonlinear Schrödinger equation (NLS) is now modified with a coupling to a mean term. In water waves similar systems were found earlier, first by Benney and Roskes [29], [see also 30, who introduced surface tension] and later by Davey and Stewarts.K [31] who put the equations into a simpler form.

One of the major difficulties in the study of water waves is the determination of the free surface, which appears as an unknown in the basic formulation of the problem. For two dimensional water waves, where the free surface evolves as a function of one space dimension and time, there are various techniques which can be used to eliminate the vertical coordinate and to reduce the problem to the evaluation of the motion of the wave height and velocity potential on the free surface. Effective methods used in the two dimensional water wave problem include conformal mapping and singular integral equations which make use of complex analytic techniques [see e.g. 32–35]. For the three dimensional problem, where the free surface evolves as a function of two space dimensions and time, the situation is more difficult and one loses the possibility of employing complex analysis.

Zakharov [36] showed that the wave height η and velocity potential ϕ evaluated on the free surface, are canonically conjugate variables and formulated the water wave equations as a Hamiltonian system. Craig and Sulem [37] employed these variables and introduced an elegant Dirichlet-Neumann operator $G(\eta)$ associated with the velocity potential which eliminated the vertical coordinate from the formulation. The operator $G(\eta)$ is obtained as a series, which is valid for small η. This formulation was used in Craig and Groves [38] to find small amplitude/long wave approximations including the Boussinesq, Korteweg-deVries (KdV) and Kadomtsev-Petviashvili (KP) equations. In Craig and Nicholls [39] this formulation was used to prove the existence of traveling periodic water waves. The Dirichlet-Neumann operator methodology, which employs high order series approximations to a modified version of G, has been employed to perform interesting computational investigations [40].

In [41] a new explicit nonlocal formulation of water waves for both 1+1 and 2+1 dimensions was constructed. The original equations with unknown boundary conditions are replaced by an integro-differential equation and a nonlinear partial differential equation, both of which are formulated in a known domain. The vertical coordinate is removed from the determining equations. These two equations can be used to determine the wave height and the velocity potential on the free surface. From this system well known asymptotic equations, in both shallow and deep water with surface tension included, are obtained and agree in the shallow water limit [38].

In this paper, we use this nonlocal system for water waves, and some of the shallow water reductions, notably the Benney-Luke (BL) system [42] and Kadomtsev-Petviashvili (KP) equation, and discuss the results of employing the SPRZ method to find lump type solutions to all of these equations.

Finally, we briefly mention some new results involving the intermediate long wave (ILW) equation in 1+1 and 2+1 dimensions which arises in a multi-fluid system [cf 43, and references therein]. The ILW equations and its "infinite depth" limit, the $1+1$ and $2+1$ Benjamin-Ono equations are novel in that they are singular integro-differential equations. We show how the SPRZ method can be applied to find localized modes of these interesting systems.

2. Dispersion-managed systems

In nonlinear fiber optics, the following normalized Nonlinear Schrödinger (NLS) type equation, sometimes called the perturbed NLS equation, plays a central role

$$iu_z + \frac{1}{2}D(z)u_{tt} + g(z)|u|^2u = 0 \tag{2.1}$$

where $D(z)$ is the normalized group velocity dipsersion, and $g(z)$ the damping and amplification; [see 1, 11, 12, for further details about the scope and applicability of these equations]. Note that ther usual time and space scales are interchanged so that now z plays the role of an evolution variable. The so called "classical" NLS equation has $D=1, g=1$ and a classical soliton solution is given by

$$u = \eta \operatorname{sech}(\eta t) e^{-i\eta^2 z}$$

In equation 2.1 we take the dispersion to be comprised of an average part and a large and rapidly varying portion.

$$D(z) = \langle d \rangle + \frac{1}{z_a} \Delta\left(\frac{z}{z_a}\right)$$

where z_a is a small parameter, $z_a = l_{\mathrm{m}}/z_* \ll 1$

$$\Delta = \begin{cases} \Delta_1 & \text{in "anomalous"} \\ \Delta_2 & \text{in "normal"} \end{cases}$$

with Δ having zero average in every map period. $\langle \Delta \rangle = \frac{\int_0^{z_a} \Delta(z) dz}{z_a} = 0$. The loss-amplification term, $g(z) = g\left(\frac{z}{z_a}\right)$, is typically given by the formula

$$g(z) = e^{-2\Gamma(z - nz_a)/z_a}, \quad nz_a < z < (n+1) z_a,$$

for $n = 1, 2, 3 \cdots$ and periodically extended outside each map period.

It is well known that the perturbed NLS equation 2.1 (PNLS) can be derived from Maxwell's equations [cf 1, 11]. We employ the following non-dimensional variables [19]:

$$z' = \frac{z}{z_*}, \quad t' = \frac{t}{t_*}, \quad u = \frac{E}{\sqrt{P_*}}, \quad z_a = \frac{l_m}{z_*}$$

where E is the electromagnetic field, and P_*, t_*, z_* are characteristic values of the power, time and nonlinear length scales. Typical values are: $P_* = 1$ mW; $t_* = 12$ ps (for a pulse full-width half-maximum, FWHM, of 20 ps), and to balance all terms in the PNLS equation we take: $z_* = \frac{1}{\gamma P_*}$ where γ is the effective nonlinear coefficient in dimensional NLS equation [cf 19]. Using $\gamma = 2.5\ \mathrm{W}^{-1}\mathrm{km}^{-1}$ leads to $z_* = 400$ km; similarly using a dimensional map length (or amplifier length) of $l_m = 40$ km we find $z_a = 0.1$. z_a is an important small parameter in the analysis.

In figure 1 we depict a typical communication system with its corresponding dispersion, $D(z)$, and damping and loss, $g(z)$, profiles. In the top portion of the figure EDFA represents amplifiers ("Erbium doped fiber amplifiers"); in most long distance systems one also has pre- and post-compensation fibers which compensate for the nonzero average fiber dispersion.

The asymptotic analysis of the PNLS equation 2.1,

$$iu_z(z,t) + \frac{1}{2}\left[\langle d \rangle + \frac{1}{z_a}\Delta\left(\frac{z}{z_a}\right)\right] u_{tt} + g\left(\frac{z}{z_a}\right) |u|^2 u = 0$$

employs the method of multiple scales [18] where

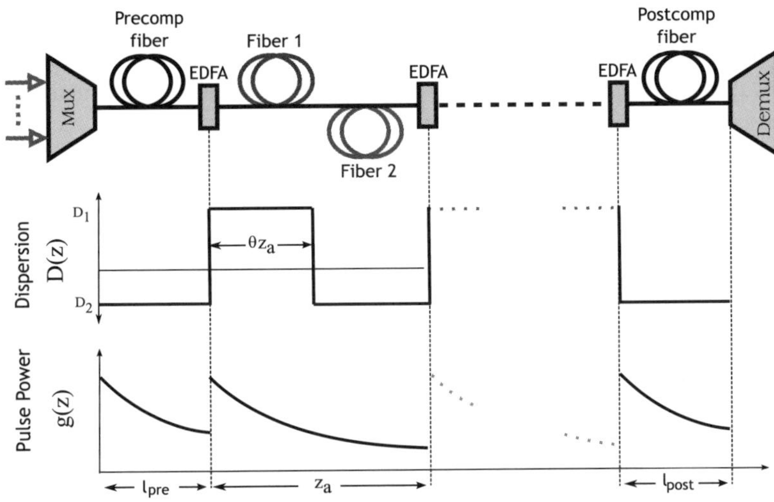

Figure 1. Dispersion-managed system: Top: fibers with pre-and post compensation; EDFA amplifiers and two pieces of fiber in each amplifier period; Middle: dispersion map; bottom: periodic damping-amplifier function: $g(z)$.

$$u = u(\zeta, Z, t; z_a): \quad \zeta = \frac{z}{z_a}, Z = z; z_a \ll 1$$

It follows that $\partial_z u = \frac{1}{z_a}\partial_\zeta u + \partial_Z u$ and we assume a perturbation expansion in powers of z_a

$$u = u^{(0)} + z_a u^{(1)} + z_a^2 u^{(2)} + \cdots$$

At leading order, $O(\frac{1}{z_a})$, a linear equation is obtained

$$iu_\zeta^{(0)} + \frac{1}{2}\Delta(\zeta)u_{tt}^{(0)} = 0 \tag{2.2}$$

which can be solved by Fourier Transforms (FT)

$$\hat{u}^{(0)}(\omega) \equiv F\{u^{(0)}(t)\} = \int_{-\infty}^{\infty} u^{(0)}(t)e^{-i\omega t}\,dt$$

Taking the FT of equation 2.2 we find:

$$i\hat{u}_\zeta^{(0)} - \frac{\omega^2}{2}\Delta(\zeta)\hat{u}^{(0)} = 0$$

which implies

$$\hat{u}^{(0)} = \hat{U}(Z,\omega)e^{-\frac{i}{2}\omega^2 C(\zeta)} + O(z_a)\,, \quad C(\zeta) \equiv \int_0^\zeta \Delta(\zeta')d\zeta'$$

where $\hat{U}(Z,\omega)$ is arbitrary. Proceeding to next order, one has to impose a secularity condition to avoid growth in the solution $u^{(1)}$; this determines $\hat{U}(Z,\omega)$ which is found to satisfy a nonlocal, nonlinear equation which is referred to as the dispersion-managed NLS (DMNLS)equation, which is given by

$$i\hat{U}_Z - \frac{1}{2}\omega^2\langle d\rangle\hat{U} + \langle ge^{\frac{i}{2}\omega^2 C}F\{|u^{(0)}|^2u^{(0)}\}\rangle = 0 \tag{2.3}$$

where the average is defined as $\langle K\rangle = \int_0^1 K(\zeta)d\zeta$. Alternatively the averaged nonlinear term in the DMNLS equation can be written:

$$\langle\cdot\rangle = \iint r(\omega_1\omega_2)\hat{U}(\omega+\omega_1)\hat{U}(\omega+\omega_2)\hat{U}^*(\omega+\omega_1+\omega_2)d\omega_1 d\omega_2$$

where $r(x)(2\pi)^2 = \langle ge^{iCx}\rangle$; If $\Delta \to 0$ the DMNLS equation reduces to the classical NLS equation written in the Fourier domain. In the lossless case: $g = 1$: the kernel $r(x)$ simplifies to

$$r(x) = \frac{\sin(sx)}{(2\pi)^2 sx}$$

where $s = \frac{\theta\Delta_1}{2}$ is called the map strength; s is proportional to the area under the the curve $\Delta(\zeta)$ in one segment of the map.

One can look for DM solitons by letting: $\hat{U}(z,\omega) = \hat{f}(\omega)e^{i\lambda^2 z/2}$ Then the DMNLS equation reduces to

$$-\frac{\lambda^2}{2}\hat{f} - \frac{\omega^2}{2}\langle d\rangle\hat{f} + \langle ge^{\frac{i}{2}\omega^2 C}F\{|f(t)|^2 f(t)\}\rangle = 0 \tag{2.4}$$

or

$$-\frac{\lambda^2}{2}\hat{f} - \frac{\omega^2}{2}\langle d\rangle\hat{f} + \iint r(\omega_1\omega_2)\hat{f}\cdot\hat{f}\cdot\hat{f}^*\, d\omega_1 d\omega_2 = 0 \tag{2.5}$$

where $f\cdot\hat{f}\cdot\hat{f}^* = f(\omega+\omega_1)\hat{f}(\omega+\omega_2)\hat{f}^*(\omega+\omega_1+\omega_2)$; in equation 2.3 $f(t)$ represents the soliton profile in the time domain.

Equation 2.3 is a nonlinear fixed-point equation to solve for f. We carried out early numerical computations using Petviashvili's method in Ablowitz and Biondini [18] with $\langle d\rangle > 0$ and found soliton solutions. As indicated in the introduction, this and subsequent studies motivated our recent research into alternative numerical methods [26], which is described later in this paper. It should be noted that is usually easier to carry out numerical computations with equation 2.4 rather than 2.5. Existence of soliton solutions was proven in Zharnitsky et al. [21]. When $s = \Delta_1 = 0$ we recover classical soliton : $f(t) = \lambda\text{sech}(\lambda t)$. In figure 2 we depict both classical ($s = 0$ left) as well as dispersion managed ($s = 1$ right) solitons with $g = 1, \lambda = 1$. We note that the DM soliton "breathes" in each map period.

Importantly one can also find "dark-gray" DM solitons [44] where one has to modify the approach due to the fact that the pulses do not decay at infinity. On the otherhand one can proceed to higher orders and obtain a higher order DMNLS equation. Interestingly this higher order DMNLS equation supports multi-humped DM solitons [23].

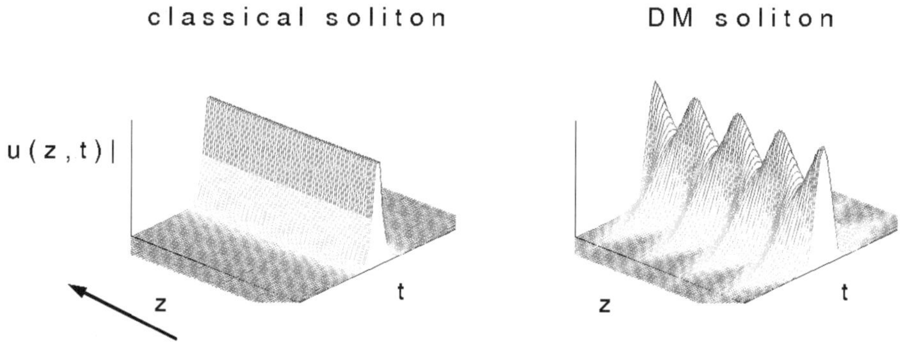

Figure 2. Classical and dispersion-managed solitons. Left: classical soliton; Right: DM soliton.

Also of interest is the fact that when $s \gg 1$ we can find useful approximate solutions to the DMNLS assuming $\hat{U}$ depends weakly on s. In [23, 28] it is shown that when $g = 1$

$$i\hat{U}_Z - \frac{1}{2}\omega^2\langle d\rangle\hat{U} + \Phi(|\hat{U}|^2)\hat{U} \sim 0 \tag{2.6}$$

where

$$\Phi(|\hat{U}|^2) = \frac{1}{2\pi s}[(\log\, s - \gamma)|\hat{U}|^2 - \int_{-\infty}^{\infty} f(\omega - \omega')|\hat{U}|^2(\omega')d\omega']$$

and $f(x) = \frac{1}{\pi}\int \log(t)e^{i\omega t}dt$. We may solve equation 2.6

$$\hat{U}(\omega, z) = \hat{U}_0 exp[-i\langle d\rangle\omega^2 z + i\Phi(|\hat{U}_0|^2)] \tag{2.7}$$

where $\hat{U}_0 = \hat{U}(\omega, 0)$. Equation 2.7 is referred to as a quasi-linear pulse. It has a weakly nonlinear dependence on the amplitude through Φ. In communications applications quasi-linear pulses are of more interest than solitons,

On the other hand recent research has shown that DM solitons find important use in the field of frequency metrology. In this application mode locked lasers, such as Ti:sapphire lasers, have been used to produce sequences of ultra-short pulses consisting of long trains of regularly spaced pulses. This situation is depicted in figure 3. In the left part of the figure pulses circulate in a cavity consisting of a Ti:sapphire crystal of normal dispersion and a prism pair with nearly compensating anomalous dispersion. Modes emanate from an output coupler into a string of regularly spaced pulses; the pulse train is depicted in the right portion of the figure. Typical values of the parameters are: FWHM: $\tau = 10$ fs$=10^{-14}$ sec, Pulse spacing: $T_{\text{rep}} = 10$ ns $= 10^{-8}$ sec with a frequency of: $f = 1/T_{\text{rep}}$. Very recent research has shown that DM solitons play an important role in these laser systems

[45, 46]. Researchers have also been investigating other laser systems, such as mode-locked Sr:Forsterite, fiber lasers. An interesting application is the development of optical clocks which have the potential to be considerably more accurate than atomic clocks.

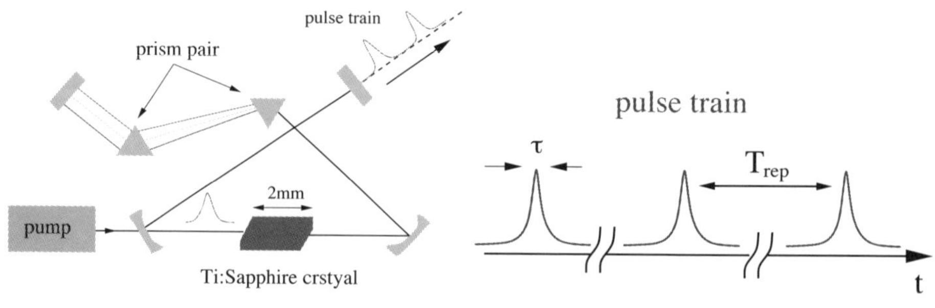

Figure 3. Left: A schematic of a Ti-Saphhire laser. Right: The pulse train from the laser.

3. NLSM systems

In this section we discuss a class of nonlinear wave equations which have focusing singularities [cf 47] A prototypical equation that arises in many physical circumstances is the $(2+1)$D focusing cubic nonlinear Schrödinger equation (**NLS**),

$$iu_z(x,y,z) + \frac{1}{2}\Delta u + |u|^2 u = 0, \qquad u(x,y,0) = u_0(x,y), \tag{3.1}$$

where u is the slowly-varying envelope of the wave, z is the direction of propagation [a], (x,y) are the transverse directions, $\Delta u \equiv u_{xx} + u_{yy}$, and u_0 is the initial conditions. In optics this equation arises in so-called Kerr media where the leading order nonlinear polarization term is cubic. Remarkably, in 1965 Kelley [48] carried out direct numerical calculations of equation (3.1) that indicated the possibility of wave collapse. In 1970 Vlasov et al. [49] proved that solutions of equation (3.1) satisfy the following "Virial Theorem" (also called Variance Identity)

$$\frac{d^2}{dz^2}\int (x^2+y^2)|u|^2 = 4H, \qquad H = \frac{1}{2}\int \left(|\nabla u_0|^2 - |u_0|^4\right), \tag{3.2}$$

where $\nabla \equiv (\partial_x, \partial_y)$, the integrations are carried over the (x,y) plane, and H, which is a constant of motion, is the Hamiltonian of equation (3.1). Using the Virial Theorem, Vlasov et al. [49] concluded that the solution of the NLS equation can become singular in finite distance (or time), because a positive-definite quantity could become negative for initial conditions satisfying $H < 0$. Subsequently, Weinstein [50] showed that when the power (which is also conserved) is sufficiently small, i.e., $N = \int |u_0|^2 = const < N_c \approx 1.8623\pi$, the solution exists globally, i.e., for all $z > 0$. Therefore, a *sufficient condition* for collapse

[a] here z plays the role of the evolution variable (i.e., like time).

is $H < 0$ while a *necessary condition* for collapse is $N > N_c$. Weinstein also found that the *ground-state* of the NLS plays an important role in the collapse theory. This ground-state is a "stationary" solution of the form $u = R(r)e^{it}$, [51] such that R is radially-symmetric, positive, and monotonically decaying. Papanicolaou et al. [52] studied the singularity structure near the collapse point and showed asymptotically and numerically that collapse occurs with a (quasi) self-similar profile. The readers are referred to Sulem and Sulem [53] for a comprehensive review of related studies. Recent research by Merle and Raphael [54] further elaborated on the collapse behavior of NLS equation (3.1) and related equations, allowing for detailed understanding of the self-similar asymptotic profile. Furthermore, Moll et al. [55] recently carried out detailed optical experiments in cubic media that reveal the nature of the singularity formation and showed experimentally that collapse occurs with a self-similar profile.

On the other hand, there are considerably fewer studies of wave collapse that arises in nonlinear media, whose governing system of equations have *quadratic nonlinearities*, such as water waves and $\chi^{(2)}$ nonlinear-optical media. Here we discuss a class of such systems, denoted as NLS-Mean (**NLSM**) systems, which are sometimes referred to as Benney-Roskes [29] or Davey-Stewartson [31] type. Broadly speaking, the derivation of NLSM systems is based on an expansion of the slowly-varying (i.e., quasi-monochromatic) wave amplitude in harmonics of the fundamental frequency, as well as a mean term that corresponds to the zero'th harmonic.

NLSM equations were originally obtained by Benney and Roskes [29] in their study of the instability of multidimensional wave packets in water of finite depth h, without surface tension. In 1974, Davey and Stewarts.K [31] studied the evolution of a 3D wave packet in water of finite depth and obtained a simpler, although equivalent, form of these equations. In 1975 Ablowitz and Haberman [56] studied the integrability of similar types of NLSM systems. These integrable systems correspond to the Benney-Roskes equations in the shallow water limit. In 1977 Djordjevic and Redekopp [30] extended the results of Benney and Roskes to include surface tension. Subsequently, Ablowitz and Segur [57] investigated these NLSM water wave systems with surface tension included. They showed that the shallow water limit, i.e., $h \to 0$, corresponds to those obtained by Ablowitz and Haberman [56]. Hence, the shallow-water limit of water wave NLSM system is integrable and can be obtained from an associated compatible linear scattering system. In Fokas and Ablowitz [58] these reduced equations were linearized by the inverse scattering transform (see also [43]). Subsequently, Ablowitz and Segur [57] studied the general NLSM system arising in water waves and showed, by a variance identity that generalizes the one for the standard NLS equation that the equations had a singularity in finite time.

More recently Ablowitz et al. [27, 28] found, from first principles, that NLSM type equations describe the evolution of the electromagnetic field in quadratically (i.e., $\chi^{(2)}$) polarized media, i.e., materials that have a *quadratic* nonlinear response. Such materials are anisotropic, e.g., crystals whose optical refraction has a preferred direction. Both scalar and vector $(3+1)$D NLS systems were obtained.

From the point of view of perturbation analysis, it is interesting to remark that in the expansion of the field in the case of water-waves the mean term in the coupled equations (ϕ in equation (3.3) below) is a velocity potential and appears as an $O(\varepsilon)$ term, whereas in the in the case of optics, the mean term is related to the electromagnetic field (ϕ_x in equation (3.3) below) appears as an $O(\varepsilon^2)$ term. However, the physically measurable quantity in water waves is the velocity which is related to derivatives of the velocity potential (e.g. ϕ_x), which scales like $O(\varepsilon^2)$, because ϕ is slowly-varying. In optics the mean term is related to the electromagnetic field and naturally appears at $O(\varepsilon^2)$. Therefore, the expansions in the water-wave and optics cases are, in fact, analogous from the viewpoint of perturbation analysis.

Thus this asymptotic analysis leads to a system of equations that describes the nonlocal-nonlinear coupling between a dynamic field that is associated with the first harmonic (with a "cascaded" effect from the second harmonic), and a static field that is associated with the mean term (i.e., the zero'th harmonic). The models considered here are special cases of both water waves and optics when the coefficients take on particular signs, and additionally in optics, we only consider the steady problem. The general NLSM system we will discuss can be written in the following non-dimensional form,

$$iu_z + \frac{1}{2}(\sigma_1 u_{xx} + u_{yy}) + \sigma_2 u|u|^2 - \rho u \phi_x = 0, \tag{3.3}$$

$$\phi_{xx} + \nu\phi_{yy} = \left(|u|^2\right)_x ,$$

where $u(x,y,t)$ corresponds to the field associated with the first-harmonic, $\phi(x,y,t)$ corresponds to the mean field, σ_1 and σ_2 are ± 1, and ν and ρ are real constants that depend on the physical parameters. It it well-known that the system (3.3) can admit collapse of localized waves when $\sigma_1 = \sigma_2 = 1$ and $\nu > 0$. In that case, the governing equations are

$$iu_z + \frac{1}{2}\Delta u + |u|^2 u - \rho u \phi_x = 0 , \tag{3.4a}$$

$$\phi_{xx} + \nu\phi_{yy} = (|u|^2)_x , \tag{3.4b}$$

where $\nu > 0$ and ρ is real, and the initial conditions are $u(x,y,0) = u_0(x,y)$, $\phi(x,y,0) = \phi_0(x,y)$, such that equation (3.4b) is satisfied at $z = 0$, i.e., $\phi_{0,xx} + \nu\phi_{0,yy} = (|u_0|^2)_x$. We will investigate the collapse dynamics in the NLSM System (3.4).

We note that the system (3.4) reduces to the classical NLS equation (3.1) when $\rho = 0$, because in that case the mean field ϕ does not couple to the harmonic field u in equation (3.4a). In addition, when $\nu = 0$ equation (3.4b) gives that $\phi_x = |u|^2$ and, therefore, equation (3.4a) reduces to a classical NLS equation (3.1) with the cubic term $(1-\rho)|u|^2 u$. In nonlinear optics problem it turns out that $\rho > 0$, whereas in water waves $\rho < 0$. In either case, i.e., when $\rho \neq 0$, the NLSM system (3.4) is a nonlocal system of equations. Indeed, since $\nu > 0$, equation (3.4b) can be solved as

$$\phi(x,y,z) = \int_{-\infty}^{\infty} G(x-x', y-y') \frac{\partial}{\partial x'} |u(x',y',z)|^2 \, dx'dy' ,$$

where $G(x,y)$ is the usual Green's function. For equation (3.4b) $G(x,y) = (4\pi)^{-1}\log(x^2 + y^2/\nu)$, which corresponds to a strongly-nonlocal function ϕ. While one might have expected the strong-nonlocality in the NLSM to arrest the collapse process, generally speaking, that is not the case for the system (3.4). Moreover, there is a striking mathematical similarity between collapse dynamics in the NLS and NLSM cases.

Two conserved quantities for the NLS equation (3.1) and NLSM system (3.4) are the power, i.e.,

$$N(u) = \int |u|^2 = N(u_0)\,, \tag{3.5}$$

where the integrations (here and below) are carried over the (x,y) plane, and the Hamiltonian, i.e.,

$$\begin{aligned} H_{\text{NLS}}(u) &= \frac{1}{2}\int |\nabla u|^2 - \frac{1}{2}\int |u|^4 = H_{\text{NLS}}(u_0), \\ H_{\text{NLSM}}(u,\phi) &= \frac{1}{2}\int |\nabla u|^2 - \frac{1}{2}\int |u|^4 + \frac{\rho}{2}\int (\phi_x^2 + \nu\phi_y^2) = H_{\text{NLSM}}(u_0,\phi_0), \end{aligned} \tag{3.6}$$

where H_{NLS} and H_{NLSM} correspond to equation (3.1) and System (3.4), respectively, and ϕ in (3.6) is obtained from equation (3.4b). In addition, the Virial Theorem holds [cf 59].

$$\frac{\partial^2}{\partial z^2}\int (x^2 + y^2)|u|^2 = 4H\,, \tag{3.7}$$

where H is the corresponding Hamiltonian, i.e., either H_{NLS} or H_{NLSM}. We are interested in the localized-decaying case, when u and ϕ vanish sufficiently rapidly at infinity to be in the the Sobolev space H_1, i.e., $\int |u|^2 + \int |\nabla u|^2 < \infty$ and similarly for ϕ. We note that within the context of the water-wave problem (i.e., $\rho < 0$), existence and well-posedness of solutions of System (3.4) were studied in Ghidaglia and Saut [60]. Singularity formation corresponds to finite-time (or finite-distance) blowup in H_1. Since the L_2 norm is conserved (3.5), blowup in H_1 amounts to $\lim_{z\to Z_c}\int |\nabla u|^2 = \infty$, where Z_c is the collapse distance. In fact, it is well-known in NLS and NLSM theories that when a singularity occurs, the peak amplitude of the wave blows-up as well, i.e. $\lim_{z\to Z_c}\max_{(x,y)}|u(x,y,z)| = \infty$.

When $H < 0$ it follows from the Virial Theorem [49] that the solution becomes singular in finite time. This gives a *sufficient condition* for collapse. On the other hand, a *necessary condition* for collapse can be obtained using the associated ground-state [cf 47] We note that the Hamiltonian, equation 3.6, is comprised of three integrals, the first of which is positive definite, the second negative definite, and, when $\nu \geq 0$, the third integral is definite with a sign that is determined by ρ. Generally speaking, NLS and NLSM theory shows that the positive-definite terms correspond to defocusing mechanisms, while the negative-definite terms correspond to focusing mechanisms. Thus, it follows that when $\rho > 0$, i.e., in the optics case, the coupling to the mean field corresponds to a self-defocusing mechanism, while when $\rho < 0$, i.e., the water-wave case, it corresponds to a self-focusing effect in addition to the the cubic term in the NLS equation (3.1). In other words, one can expect that self-focusing in the water-wave case is "easier" to attain than in the optics case.

A stationary solution of the NLSM system (3.4) is a solution of the form $u(x,y,z) = F(x,y)e^{i\lambda z}$ and $\phi(x,y,z) = G(x,y)$, where F and G are real functions and λ is a positive real number. Substituting this ansatz into equations (3.4) gives

$$-\lambda F + \frac{1}{2}\Delta F + F^3 - \rho F G_x = 0\,, \tag{3.8a}$$

$$G_{xx} + \nu G_{yy} = (F^2)_x\,. \tag{3.8b}$$

Similarly, the NLS stationary solutions, which are obtained by substituting $u = R(x,y)e^{i\lambda z}$ into the NLS equation (3.1), satisfy

$$-\lambda R + \frac{1}{2}\Delta R + R^3 = 0\,. \tag{3.9}$$

The *ground-state* of the NLS[b] can be defined as a solution in H_1 of equation (3.9) for a given λ having minimal power of all the nontrivial solutions. The existence and uniqueness of the ground state have been proven, as also the fact that it is radially-symmetric, positive, and monotonically decaying [53]. Since $R(r;\lambda) = \sqrt{\lambda}R(\sqrt{\lambda}r;1)$, it suffices to consider the case $\lambda = 1$. Furthermore, Weinstein [50] proved that the NLS ground-state is a minimizer of a Gagliardo-Nirenberg inequality that is associated with the NLS Hamiltonian.

Moreover, Weinstein proved that when $N < N_c$, the NLS solution exists globally (i.e., for all $z > 0$) in H_1. In addition, one can show that any stationary solution, in particular the ground-state, admits a zero Hamiltonian, i.e., $H_{\mathrm{NLS}}(R) = 0$. These results explain why the ground-state may be considered to be on the borderline between existence and collapse. Indeed, consider the initial conditions $u_0 = (1+\varepsilon)R(r)$ with $\varepsilon = const.$ When $\varepsilon < 0$ then $N < N_c$ and, therefore, the solution exists globally. On the other hand, when $\varepsilon > 0$ then $H < 0$ and, therefore, finite-distance collapse is guaranteed by the Virial Theorem [50]. We note that $N \geq N_c$ is only a necessary condition for collapse, i.e., there are solutions with $N > N_c$ that exist globally.

Similarly to the NLS case, the ground-state of the NLSM system (3.8) can be defined as the nontrivial solution (F,G) in H_1, such that F has minimal power. Cipolatti [61] proved the existence of the ground-state and in the same spirit as for the NLS, Papanicolaou et al. [62] showed that the ground-state is a minimizer of a suitable functional and that there is critical power N_c such that when the initial data of the NLSM system satisfies $\int |u_0|^2 < N_c$ the solution of equations (3.4) exists in H_1 for all $z > 0$. In other words, solutions of the NLSM system (3.4) exist globally when their power is smaller than the power of the corresponding ground-state.

Thus it is important to be able to understand the features of the ground state of the system (3.8) and to be able to compute them. We use the numerical methods described in this paper to obtain the results described below.

It is convenient to define the astigmatism of the ground-state via

$$e(F) \equiv \frac{\int |(F^2)_x|}{\int |(F^2)_y|}\,. \tag{3.10}$$

[b] R, the NLS ground-state, is sometimes referred to as the Townes profile.

It follows from equation (3.10) that $e = 1$ corresponds to a radially-symmetric ground-state, and $e < 1$ and $e > 1$ correspond to a ground-state that is relatively wider along the x and y axes, respectively. In other words, $e \approx L_y/L_x$, where L_x and L_y are the full-widths at half-max of the function.

Figure 4(a,b) shows the on-axes amplitudes of the ground-state for $\rho = 0$ (i.e., the radially-symmetric R profile); $(\nu, \rho) = (0.5, -1)$; and $(\nu, \rho) = (0.5, 1)$. The contour plots in figure 4(c) and (d) correspond to the $\rho = -1$ and $\rho = 1$ cases, respectively. These plots clearly show that the ground-states with $\rho \neq 0$ are astigmatic.

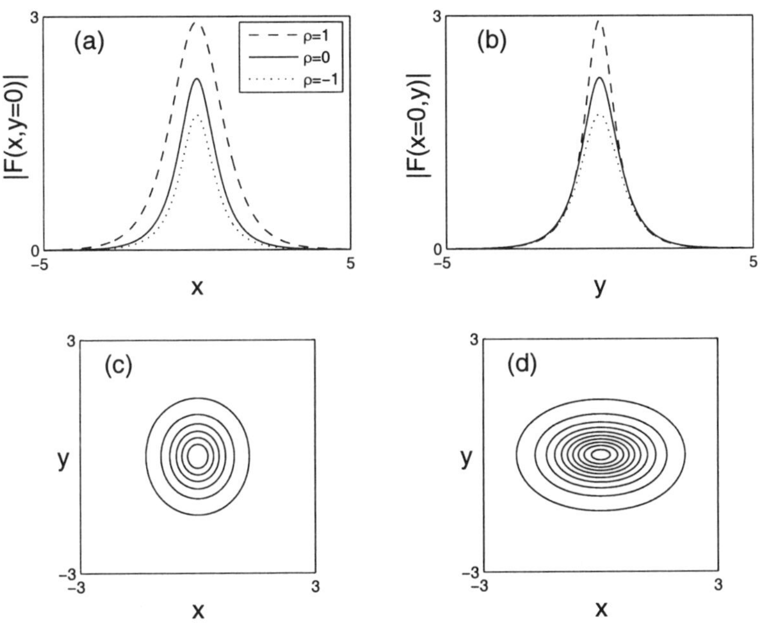

Figure 4. Top: the on-axes amplitudes of the ground-state (a) along the y-axis and (b) along the x-axis for $(\nu, \rho) = (0.5, 1)$ (dashes), $\rho = 0$ (solid), and $(\nu, \rho) = (0.5, -1)$ (dotted). Bottom: contour plots of $F(x, y)$ for: (c) $\rho = -1$ (corresponding to dotted above) with astigmatism [i.e. equation (3.10)] $e \approx 1.17$; (d) $\rho = 1$ (corresponding to dashes above) with $e \approx 0.59$.

Asymptotic analysis and numerical simulations strongly suggest that when collapse occurs in NLS equation (3.1), under quite general conditions, it occurs with a quasi self-similar profile that is a modulation (up to a phase) of the ground-state [cf 53], i.e.,

$$|u(x, y, z)| \sim \frac{1}{L(z)} R(\frac{r}{L(z)}) , \tag{3.11}$$

where (x, y) are in some region surrounding of the collapse point (which typically shrinks during the self-focusing process), $R(r)$ is the NLS ground-state and $L(z)$ is a modulation function, such that $\lim_{z \to Z_c} L(z) = 0$, where Z_c is the collapse distance (or time). In the NLS case, the ground-state $R(r)$ is radially-symmetric and NLS-collapse simulations have shown that collapse occurs with a radially-symmetric profile. The quasi self-similar collapse has received much theoretical attention [cf 63]. However, it is very difficult to justify equation (3.11) rigorously. Only recently did Merle and Raphael [54] provide a sharp result explaining this quasi self-similar behavior in the case of the NLS equation (3.1). On the experimental side, Moll et al. [55] recently carried out detailed measurements in optical Kerr media showing that the collapse process occurs with a self-similar profile, in consistency with equation (3.11).

In contrast to the NLS case, as we have seen above, the NLSM System (3.4) when $\rho \neq 0$ and $\nu > 0$ is not rotationally invariant and the stationary solutions of (3.8) are not radially symmetric. Moreover, with this choice of parameters the stationary solutions cannot be transformed into radially-symmetric functions by any rescaling of x and y. Therefore, the NLSM ground-state, $F(x,y)$, is inherently astigmatic, which makes the analysis and numerical simulations more difficult. The asymptotic analysis of Papanicolaou et al. [62] indicates that, similar to the NLS collapse, NLSM collapse occurs with a modulated profile, i.e.,

$$|u(x,y,z)| \sim \frac{1}{L(z)} P\left[\frac{x}{L(z)}, \frac{y}{L(z)}, b(z)\right], \tag{3.12}$$

for certain functions $P(x,y,z)$, $L(z)$, and $b(z)$, such that as $z \longrightarrow Z_c$, $L(z)$ and $b(z)$ approach zero and $P(x,y,z)$ asymptotically approaches the corresponding ground-state $F(x,y)$. Numerical simulations of the NLSM using "dynamic rescaling" suggested that, indeed, the collapsing solution approaches a modulated profile. However, in Papanicolaou et al. [62] the ground-state itself was not computed. Since it was not computed, it could not be shown (numerically) that the asymptotic profile approaches the corresponding ground-state. The numerical results in this section suggest that, down to moderately small values of $L(z)$, the amplitude of the collapsing solution behaves as

$$|u(x,y,z)| \sim \frac{1}{L(z)} F\left[\frac{x}{L(z)}, \frac{y}{L(z)}\right], \tag{3.13}$$

where $F(x,y)$ is the ground-state of equations (3.4). Therefore, the results of our studies strengthen those of Papanicolaou et al. [62], because the collapsing wave is directly compared to the corresponding ground-state and is shown to approach a quasi self-similar modulation of the ground-state itself.

NLSM collapse is studied numerically by solving equations (3.4) with Gaussian initial conditions:

$$u_0^G(x,y) = \sqrt{\frac{2N}{\pi}}\, e^{-(x^2+y^2)}\,, \tag{3.14}$$

where $N = N(G)$ is the input power of u_0^G. The input power for these calculations is taken as $1.2N_c(\nu = 0.5, \rho = 1) \approx 12.2$. We note that this value of N_c is approximately twice as large as $N_c(R)$ and approximately 3.3 times larger than $N_c(\nu = 0.5, \rho = -1)$. The self-focusing dynamics and quasi-self similar behavior are understood from the simulations

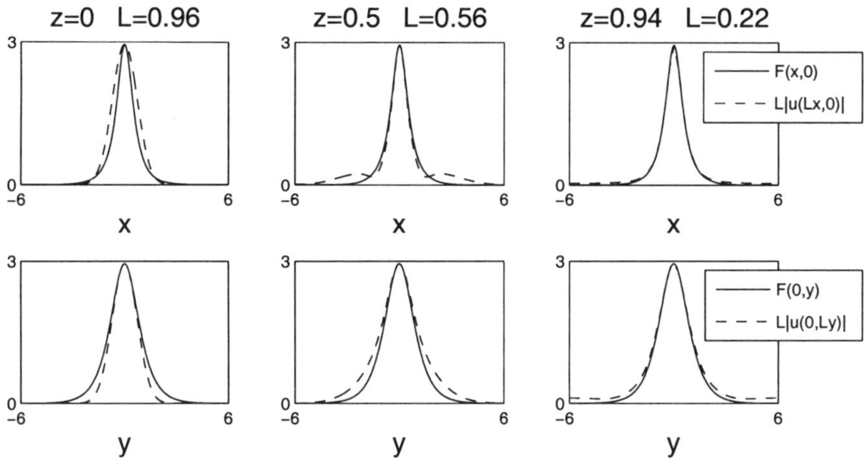

Figure 5. Convergence of the modulated collapse profile (dashes) to the NLSM ground state (solid) along the x axis (top) and the y axis (bottom) with $(\nu,\rho) = (0.5,1)$. The initial conditions are (3.14) with $N = 1.2N_c(\nu,\rho)$.

using a "modulation function" which in turn is recovered from the solution,

$$L(z) = \frac{F(0,0)}{|u(0,0,z)|},$$

where where $F(x,y)$ is the corresponding ground-state; note that $L(z)$ is a function of the propagation distance z.

The rescaled amplitude of the solution of the NLSM, i.e., $L|u(L\tilde{x},L\tilde{y},z)|$, is compared with $F(\tilde{x},\tilde{y})$, where $F(\tilde{x},\tilde{y})$ is the ground-state and $(\tilde{x},\tilde{y}) = (\frac{x}{L},\frac{y}{L})$. In order to show that the collapse process is, indeed, quasi self-similar with the corresponding ground-state, the rescaled amplitude is shown to converge pointwise to F near the origin as $z \longrightarrow Z_c$ (i.e., near the collapse point).

Figure 5 shows that the NLSM collapse is indeed self-similar with the ground-state for $\nu = 0.5$ and $\rho = 1$. The rescaled on-axis amplitude is compared separately on the x and y axes (top and bottom plots, respectively). One can see that, as the solution is undergoing self-focusing [i.e, as $L(z)$ approached zero], its rescaled profile approaches that of the astigmatic ground-state near the origin.

While the spatial region in the vicinity of the collapse point is self-similar to the ground-state, the outer "wings" of the solution do not approach the ground-state. Since the input power is approximately 20% above N_c, the residual 20% radiates into the outer wings in a process that is not self-similar with the ground-state.

Thus nonlinear-wave systems that admit a quadratic-cubic type interaction, such as in nonlinear optics and in nonlinear free-surface water waves, lead to the NLSM system (3.4). The NLSM system can admit finite-distance collapse in a certain parameter regime. The regions of collapse and global-existence can be explored in terms of the critical power, Virial Theorem, and numerical simulations of the NLSM system (3.4). Numerical simulations of the NLSM show that the collapse process occurs with a quasi self-similar profile, which is a modulation of the ground-state profile. The ground-state profile is found using the numerical methods described in this paper. Generically, the ground-state profile is astigmatic and, therefore, the collapse profile is astigmatic as well. These results are in the same spirit as for the NLS equation (3.1). However, NLSM theory is more difficult and currently not as advanced as NLS theory. From the experimental perspective, self-similar collapse in quadratic-cubic type media remains an interesting problem to be demonstrated in either free-surface water waves with surface tension or nonlinear optics.

4. Spectral renormalization

Optical temporally or spatially localized solutions in nonlinear media have attracted considerable attention in the scientific community. They have been found to exist in a wide range of physical systems, some of which have been discussed earlier. A central issue for these types of nonlinear localized waves is how to compute localized, i.e. soliton, solutions which generally involve solving nonlinear ordinary or partial differential equations. To date, various techniques have been used; e.g. shooting and relaxation techniques and methods utilizing the important concept that a soliton forms when the optical field induces a waveguide structure (or self-induced potential well) via the nonlinearity and "self-traps" [64].

Another method, first introduced by Petviashvili [24], to construct localized solutions of a nonlinear system is based on transforming to Fourier space and determining a "convergence factor" based upon the homogeneity of the nonlinearity. While it was first used to find localized solutions in the two-dimensional Korteweg-deVries equation (usually referred to as the Kadomtsev-Petviashvili equation [cf 59]), the method has been significantly extended and has been used to find localized solutions in a wide variety of interesting systems — e.g. dispersion-managed and diffraction managed (i.e. discrete systems) nonlinear Scrödinger (NLS) equations [18, 65] dark and gray soltions [44]. However, this method often is only successful when the underlying equation has a fixed nonlinearity; i.e. fixed homogeneity. However, many physically interesting problems involve nonlinearities with different homogeneities.

Below we describe a novel numerical scheme in order to compute localized solutions in nonlinear waveguides [26]. The essence of the method is to: i) transform the underlying equation governing the soliton into Fourier space (this part is the same as Petviashvili [24]); ii) re-normalize variables and iii) determine an algebraic system which is coupled to a nonlinear integral equation. Thus, we have a nonlinear nonlocal integral equation (or system of integral equations) coupled to an algebraic equation (or system). The coupling is found to prevent the numerical scheme from diverging. We have found the method of coupling to be effective and straight forward to implement. The localized pulse is determined from a convergent fixed point iteration scheme.

We describe the method using a scalar nonlinear Schrödinger like equation

$$i\frac{\partial U}{\partial z} + \nabla^2 U + V(\mathbf{x})U + N\Big(|U|^2, a(\mathbf{x})\Big)U = 0, \tag{4.1}$$

where z is the propagation direction; N is the nonlinearity that can depend on both intensity and inhomogenieties $a(\mathbf{x})$; e.g. $V(\mathbf{x})$ can model an optical lattice. Here, $\nabla^2 = \frac{\partial^2}{\partial x^2} + \frac{\partial^2}{\partial y^2}$. A special class of soliton solution can be constructed by assuming $U(\mathbf{x},z) = u(\mathbf{x};\mu)e^{i\mu z}$ where μ is the propagation constant or the soliton eigenvalue. Substituting the above ansatz into equation (4.1) we get

$$-\mu u + \nabla^2 u + V(\mathbf{x})u + N\Big(|u|^2, a(\mathbf{x})\Big)u = 0\,. \tag{4.2}$$

This is a nonlinear eigenvalue problem for u and μ which is suplemented with the following boundary conditions:

$$u \longrightarrow 0 \text{ as } |r| \longrightarrow +\infty$$

where $r^2 = x^2 + y^2$. The spectral renormalization (SPRZ) scheme is based on Fourier analysis which transforms equation (4.2) into nonlocal equation which will then be solved using a convergent scheme. First we define the Fourier transform $\mathcal{F}$ and its inverse $\mathcal{F}^{-1}$

$$\begin{aligned}\hat{u}(\mathbf{k}) &= \mathcal{F}[u(\mathbf{x})] \\ &= \int_{-\infty}^{+\infty}\int_{-\infty}^{+\infty} u(\mathbf{x})e^{-i(k_x x + k_y y)}d\mathbf{x}\,,\end{aligned} \tag{4.3}$$

$$\begin{aligned}u(\mathbf{x}) &= \mathcal{F}^{-1}[\hat{u}(\mathbf{k})] \\ &= \frac{1}{(2\pi)^2}\int_{-\infty}^{+\infty}\int_{-\infty}^{+\infty} \hat{u}(\mathbf{k})e^{+i(k_x x + k_y y)}d\mathbf{k}\,,\end{aligned} \tag{4.4}$$

where $d\mathbf{x} = dxdy$ and $d\mathbf{k} = dk_x dk_y$. Applying the Fourier transform on equation (4.2) leads to

$$\hat{u}(\mathbf{k}) = \frac{\mathcal{F}[Vu] + \mathcal{F}[N(|u|^2, a(\mathbf{x}))u]}{\mu + |\mathbf{k}|^2}\,. \tag{4.5}$$

The idea underlying this method is to construct a condition which limits the amplitude under iteration from either growing without bound or tending to zero. This is accomplished by introducing a new field variable (i.e. renormalizing the field variable)

$$u(\mathbf{x}) = \lambda w(\mathbf{x})\,, \quad \hat{u}(\mathbf{k}) = \lambda\hat{w}(\mathbf{k})\,, \tag{4.6}$$

where $\lambda \neq 0$ is a constant to be determined. Then function $\hat{w}$ satisfies

$$\hat{w}(\mathbf{k}) = \frac{1}{\lambda}\frac{\mathcal{F}[\lambda V w] + \mathcal{F}[N(\lambda^2|w|^2, a(x))w]}{\mu + |\mathbf{k}|^2} \equiv Q_\lambda[\hat{w}(\mathbf{k})]\,. \tag{4.7}$$

Multiplying equation (4.7) by $\hat{w}^*(\mathbf{k})$ and integrating over the entire (k_x, k_y) space we find the relation

$$\int_{-\infty}^{+\infty} |\hat{w}(\mathbf{k})|^2 d\mathbf{k} = \int_{-\infty}^{+\infty} \hat{w}^*(\mathbf{k})Q_\lambda[\hat{w}(\mathbf{k})]d\mathbf{k}\,. \tag{4.8}$$

Equation (4.8) provides an algebraic condition on the constant λ which, in general we denote by

$$G(\lambda) = 0\,. \tag{4.9}$$

To obtain the desired solution, we iterate Eqs. (4.7) and (4.9) as follows:

$$\hat{w}_{m+1}(\mathbf{k}) = \frac{1}{\lambda_m} \frac{\mathcal{F}[\lambda_m V w_m] + \mathcal{F}[N(\lambda_m^2 |w_m|^2, a(x))\lambda_m w_m]}{\mu + |\mathbf{k}|^2} . \tag{4.10}$$

$$G(\lambda_m) = 0 . \tag{4.11}$$

Note that it is possible that the algebraic equation (4.11) can admit more than one "root" or even complex solutions. In that case, one might need to exclude some solutions depending upon the physics at hand. Knowing the weakly nonlinear limit is very useful in this regard.

Thus the idea behind the method is to transform the underlying equation governing the localized mode, such as a nonlinear Scrödinger-type equation, into Fourier space, renormalize variables and then determine a nonlinear nonlocal integral equation coupled to an algebraic equation. The coupling is found to prevent the numerical scheme from diverging. The nonlinear guided mode is then obtained from a convergent fixed point iteration scheme. This method has already found wide applications in nonlinear optics, water waves, internal waves and related fields such as Bose-Einstein condensation [66].

5. A Nonlocal Formulation of Water Waves in Three Dimensions

In Ablowitz et al. [41] a nonlocal formulation of water waves is developed and lump solutions of the water wave equations are obtained by the SPRZ method. Here we will discuss some of these results. We begin by considering the classical gravity water wave problem without surface tension. Let us define the domain D by

$$D = \{-\infty < x_j < \infty, \quad j = 1,2, \quad -h < y < \eta(x_1,x_2,t), \quad t > 0\},$$

where $h > 0$. The water wave equations satisfy the following system for $\phi(x_1,x_2,y,t)$ and $\eta(x_1,x_2,t)$:

$$\Delta\phi = 0 \quad \text{in} \quad D \tag{5.1}$$

$$\phi_y = 0 \ \text{ on } \ y = -h \tag{5.2}$$

$$\eta_t + \phi_{x_1}\eta_{x_1} + \phi_{x_2}\eta_{x_2} = \phi_y \quad \text{on} \quad y = \eta \tag{5.3}$$

$$\phi_t + \frac{1}{2}|\nabla\phi|^2 + g\eta = \frac{\sigma}{\rho}\nabla\cdot\left(\frac{\nabla\eta}{\sqrt{1+|\nabla\eta|^2}}\right) \quad \text{on} \quad y = \eta, \tag{5.4}$$

where g is gravity, σ and ρ denote the constant surface tension and density respectively and h is the constant unperturbed fluid depth. Equation (5.3), the kinematic condition, implies that fluid particles on the free surface remain on the free surface, whereas equation

(5.4), the so-called dynamic boundary condition or Bernoulli's equation, implies continuity of pressure across the free surface. Equation (5.4) describes the dynamics of the velocity potential on the free surface. We assume that η as well as the *derivatives* of ϕ vanish as $x_1^2+x_2^2 \to \infty$.

In what follows, we will reformulate these equations in terms of the two functions η and q, where q is the value of ϕ on the free boundary [see 36], where these variables were introduced and shown to be canonically conjugate coordinates),

$$q(x_1,x_2,t) = \phi(x_1,x_2,\eta(x_1,x_2,t),t).$$

The definition of q as well as equation (5.3) can be used to express the spatial derivatives of ϕ on the boundary, in terms of η and q. Indeed, differentiating equation (5) with respect to x_1 and x_2, we find

$$\phi_{x_1} + \phi_y \eta_{x_1} = q_{x_1}, \tag{5.5a}$$

$$\phi_{x_2} + \phi_y \eta_{x_2} = q_{x_2}. \tag{5.5b}$$

Solving equations (5.5) for $\phi_{x_1}, \phi_{x_2}, \phi_y$ in terms of q we find

$$\phi_{x_1} = \frac{(1+\eta_{x_2}^2)q_{x_1} - \eta_{x_1}\eta_{x_2}q_{x_2} - \eta_{x_1}\eta_t}{1+|\nabla\eta|^2} \tag{5.6a}$$

$$\phi_{x_2} = \frac{(1+\eta_{x_1}^2)q_{x_2} - \eta_{x_1}\eta_{x_2}q_{x_1} - \eta_{x_2}\eta_t}{1+|\nabla\eta|^2} \tag{5.6b}$$

$$\phi_y = \frac{\eta_t + \eta_{x_1}q_{x_1} + \eta_{x_2}q_{x_2}}{1+|\nabla\eta|^2} \tag{5.6c}$$

5.1. *The Dynamic Boundary Condition — Bernoulli's Equation*

Equations 5.6 imply,

$$|\nabla\phi|^2 = \frac{(1+\eta_{x_2}^2)q_{x_1}^2 + (1+\eta_{x_1}^2)q_{x_2}^2 + \eta_t^2 - 2\eta_{x_1}\eta_{x_2}q_{x_1}q_{x_2}}{1+|\nabla\eta|^2}.$$

Substituting $|\nabla\phi|^2$ in equation (5.4) and also replacing ϕ_t by $q_t - \eta_t\phi_y$, where ϕ_y is given by equation (5.6c), we find an equation involving q and η. Simplifying this equation we find

$$q_t + \frac{1}{2}|\nabla q|^2 + g\eta - \frac{(\eta_t + \nabla q\cdot\nabla\eta)^2}{2(1+|\nabla\eta|^2)} = \frac{\sigma}{\rho}\nabla\cdot\left(\frac{\nabla\eta}{\sqrt{1+|\nabla\eta|^2}}\right), \tag{5.7}$$

which yields the following equation, which we denote as equation (II):

$$q_t + \frac{1}{2}|\nabla q|^2 + g\eta - \frac{(\eta_t + \nabla q\cdot\nabla\eta)^2}{2(1+|\nabla\eta|^2)} = \frac{\sigma}{\rho}\nabla\cdot\left(\frac{\nabla\eta}{\sqrt{1+|\nabla\eta|^2}}\right) \tag{II}$$

5.2. *The Analysis of Equations (5.1) — (5.4)*

Let both functions $\phi(x_1,x_2,y)$ and $\psi(x_1,x_2,y)$ satisfy Laplace's equation (5.1). Then

$$\phi_y(\psi_{x_1x_1}+\psi_{x_2x_2}+\psi_{yy})+\psi_y(\phi_{x_1x_1}+\phi_{x_2x_2}+\phi_{yy})=0.$$

This identity can be rewritten in the following form,

$$(\phi_y\psi_{x_1}+\psi_y\phi_{x_1})_{x_1}+(\phi_y\psi_{x_2}+\psi_y\phi_{x_2})_{x_2}+(\phi_y\psi_y-\phi_{x_1}\psi_{x_1}-\phi_{x_2}\psi_{x_2})_y=0. \quad (5.8)$$

We choose for ψ the particular solution

$$E(x_1,x_2,k_1,k_2)=e^{ik_1x_1+ik_2x_2+ky}, \quad k=(k_1^2+k_2^2)^{1/2}.$$

Using this solution in equation (5.8), we find the identity

$$(E(ik_1\phi_y+k\phi_{x_1}))_{x_1}+(E(ik_2\phi_y+k\phi_{x_2}))_{x_2}+(E(k\phi_y-ik_1\phi_{x_1}-ik_2\phi_{x_2}))_y=0.$$

Then from the divergence theorem, this equation implies the *global relation*

$$\int_{\partial D} E\Big\{(ik_1\phi_y+k\phi_{x_1})N_1+(ik_2\phi_y+k\phi_{x_2})N_2+(k\phi_y-ik_1\phi_{x_1}-ik_2\phi_{x_2})N_3\Big\}dS=0, \quad (5.9)$$

where $(N_1,N_2,N_3)^T$ is a vector normal to the surface ∂D, and dS is a surface element. On the bottom:

$$N_1=N_2=0, \quad N_3=-1, \quad y=-h, \quad \phi_y=0,$$

thus the integrand on the bottom in 5.9 becomes

$$e^{ik_1x_1+ik_2x_2-kh}(ik_1\phi_{x_1}+ik_2\phi_{x_2}).$$

On the free surface:

$$N_1=-\eta_{x_1}, \quad N_2=-\eta_{x_2}, \quad N_3=1, \quad y=\eta,$$

thus the integrand on the free surface in (5.9) simplifies to:

$$e^{ik_1x_1+ik_2x_2+k\eta}[k(\phi_y-\phi_{x_1}\eta_{x_1}-\phi_{x_2}\eta_{x_2})-ik_1(\phi_{x_1}+\phi_y\eta_{x_1})-ik_2(\phi_{x_2}+\phi_y\eta_{x_2})].$$

The three parentheses appearing in the above expression equal η_t, q_{x_1}, q_{x_2}, respectively. Thus, with decaying conditions for $|\nabla\phi|$ on the sides, equation (5.9) becomes

$$\int_{-\infty}^{\infty}\int_{-\infty}^{\infty} dx_1dx_2e^{ik_1x_1+ik_2x_2}\Big\{e^{k\eta}[k\eta_t-ik_1q_{x_1}-ik_2q_{x_2}]$$

$$+e^{-kh}[ik_1\phi_{x_1}+ik_2\phi_{x_2}]\Big\}(x_1,x_2,-h,t)=0. \quad (5.10)$$

This equation is valid for both roots $\pm\sqrt{k_1^2+k_2^2}$. We can eliminate the terms at $y=-h$ from equation 5.10 by replacing k with $-k$ in equation (5.10). Multiplying by e^{kh} and subtracting the two equations with $\pm k$ respectively yields the following equation, which we denote as equation (I),

$$\int_{-\infty}^{\infty}\int_{-\infty}^{\infty} dx_1 dx_2 e^{ik_1x_1+ik_2x_2}\{i\eta_t \cosh[k(\eta+h)] + (\overrightarrow{k}\cdot\nabla q)\frac{\sinh[k(\eta+h)]}{k}\} = 0, \quad \text{(I)}$$

where k_1, $k_2 \in \mathbb{R}$ and $\overrightarrow{k} = (k_1,k_2)$ contains "Fourier-like" parameters.

In $1+1$ dimensions the nonlocal equation (I) reduces to the convenient form

$$\int_{-\infty}^{\infty} dx e^{ikx}\{i\eta_t \cosh k[\eta+h] + q_x \sinh k[\eta+h]\} = 0,\ k\in\mathbb{R}. \quad \text{(I')}$$

Thus the nonlocal equation satisfies the Laplace and kinematic equations as well as the bottom boundary conditions. Equations (I) and (II) are the two basic equations for the wave height $\eta=\eta(x_1,x_2,t)$ and the the velocity potential on the free surface $q=\phi(x_1,x_2,\eta,t)$. We note that the vertical coordinate y is removed in this formulation, and the unknown boundary η is completely determined by the nonlinear integro-differential equation (I) and the nonlinear PDE (II). Since the nonlocal equation (I) or (I') are valid for all k_1, k_2 real and k real respectively, these equations provide an integral equation formulation of the Dirichlet to Neumann map which represent the summation of the series analyzed in Craig and Sulem [37]. We also stress the explicit, spectral and relatively simple form of the free surface equation which is helpful in calculations mentioned below.

5.3. *The Linear Limit*

In the linear limit, $|\eta|\ll 1$, $|q_t|\ll 1$, $|q_{x_j}|\ll 1$, $j=1,2$, thus equations (I) and (II) become,

$$\int_{-\infty}^{\infty}\int_{-\infty}^{\infty} dx_1 dx_2 e^{ik_1x_1+ik_2x_2}\left\{ i\eta_t \cosh[kh] + \left[\left(\frac{k_1}{k}\right)q_{x_1} + \left(\frac{k_2}{k}\right)q_{x_2}\right]\sinh[kh]\right\} = 0 \quad (5.11a)$$

$$q_t + g\eta = \frac{\sigma}{\rho}(\eta_{x_1x_1}+\eta_{x_2x_2}) \quad (5.11b)$$

We define the Fourier transforms of η by

$$\hat{\eta} = \int_{-\infty}^{\infty}\int_{-\infty}^{\infty} dx_1 dx_2 e^{ik_1x_1+ik_2x_2}\eta(x_1,x_2,t),$$

and similarly for the derivatives of q.

Using these definitions, equations (5.11) yield

$$i\hat{\eta}_t + \left[\left(\frac{k_1}{k}\right)(\widehat{q_{x_1}}) + \left(\frac{k_2}{k}\right)(\widehat{q_{x_2}})\right]\tanh[kh] = 0, \quad (5.12a)$$

$$\hat{q}_t + g\hat{\eta} = \frac{\sigma}{\rho}\left[(\widehat{\eta_{x_1x_2}}) + (\widehat{\eta_{x_2x_2}})\right]. \quad (5.12b)$$

The definition of the FT implies $\widehat{(\eta_{x_1})} = -ik_1\hat{\eta}$, $\widehat{(\eta_{x_2})} = -ik_2\hat{\eta}$, etc. Differentiating equation (5.12a) with respect to t and using equation (5.12b) to express $\widehat{q_{x_1t}}$ and $\widehat{q_{x_2t}}$ in terms of $\hat{\eta}$, i.e., using

$$\widehat{q_{tx_j}} = ik_j\left(g + \frac{\sigma}{\rho}k^2\right)\hat{\eta}, \quad j = 1,2,$$

we find

$$\hat{\eta}_{tt} + \left[kg\left(1 + \frac{\sigma}{g\rho}k^2\right)\tanh[kh]\right]\hat{\eta} = 0. \tag{5.13}$$

5.4. *The Nondimensional Form*

It is convenient to have a nondimensional formulation. In this respect, we first replace in equations (I) & (II) all variables by prime variables and then we make the following substitutions

$$x_1' = lx_1, \quad x_2' = \frac{l}{\gamma}x_2, \quad k_1' = \frac{k_1}{l}, \quad k_2' = \gamma\frac{k_2}{l}, \quad t' = \frac{lt}{c_0}, \quad q' = \frac{gla}{c_0}q, \quad \eta' = a\eta,$$

where $c_0 = \sqrt{gh}$ and l, $\frac{l}{\gamma}$ are typical length scales (e.g. wavelengths) in the x_1, x_2 directions respectively. Defining the dimensionless parameters

$$\varepsilon = \frac{a}{h}, \quad \mu = \frac{h}{l},$$

equations (I) & (II) yield the following equations:

$$\int_{-\infty}^{\infty}\int_{-\infty}^{\infty} dx_1 dx_2 e^{i(k_1x_1+k_2x_2)}\Big\{i\eta_t\cosh[k\mu(1+\varepsilon\eta)] + \sinh[k\mu(1+\varepsilon\eta)]\left[\frac{k_1}{k\mu}q_{x_1} + \frac{\gamma^2k_2}{k\mu}q_{x_2}\right]\Big\} = 0 \tag{5.14}$$

$$q_t + \eta + \frac{\varepsilon}{2}(q_{x_1}^2 + \gamma^2 q_{x_2}^2) - \frac{1}{2}\varepsilon\mu^2\frac{[(\eta_t + \varepsilon(q_{x_1}\eta_{x_1} + \gamma^2 q_{x_2}\eta_{x_2})]^2}{1 + (\varepsilon\mu)^2(\eta_{x_1}^2 + \gamma^2\eta_{x_2}^2)} = \frac{\tilde{\sigma}\left[\mu^2\eta_{x_1x_1}(1 + \varepsilon^2\mu^2\gamma^2\eta_{x_2}^2) + \mu^2\gamma^2\eta_{x_2x_2}(1 + \varepsilon^2\mu^2\eta_{x_1}^2) - 2\varepsilon^2\mu^4\gamma^2\eta_{x_1}\eta_{x_2}\eta_{x_1x_2}\right]}{\left[1 + (\varepsilon\mu)^2(\eta_{x_1}^2 + \gamma^2\eta_{x_2}^2)\right]^{3/2}}, \tag{5.15}$$

where,

$$k = \sqrt{k_1^2 + \gamma^2 k_2^2}, \quad \tilde{\sigma} = \frac{\sigma}{\rho g h^2}.$$

By adding/subtracting the linear terms and by taking the Fourier transform, we can rewrite equations (5.14) and (5.15) in an alternative form in which the linear terms are separated. This is useful for various purposes, including for numerical evaluations (see the section on computational investigations):

$$i\cosh[k\mu]\hat{\eta}_t + \sinh[k\mu]\left[\left(\frac{k_1}{k\mu}\right)(\widehat{q_{x_1}}) + \left(\frac{\gamma^2 k_2}{k\mu}\right)(\widehat{q_{x_2}})\right] =$$
$$\int_{-\infty}^{\infty}\int_{-\infty}^{\infty} dx_1 dx_2 e^{i(k_1x_1+k_2x_2)}\left\{i\eta_t(cosh[k\mu] - \cosh[k\mu(1+\varepsilon\eta)])\right.$$
$$\left. + (\sinh[k\mu] - \sinh[k\mu(1+\varepsilon\eta)])\left[\frac{k_1}{k\mu}q_{x_1} + \frac{\gamma^2 k_2}{k\mu}q_{x_2}\right]\right\}, \quad (5.16)$$

$$\hat{q}_t + (1+\tilde{\sigma}\mu^2k^2)\hat{\eta} = \int_{-\infty}^{\infty}\int_{-\infty}^{\infty} dx_1 dx_2 e^{i(k_1x_1+k_2x_2)}\left\{-\frac{\varepsilon}{2}(q_{x_1}^2 + \gamma^2 q_{x_2}^2)\right.$$
$$+\frac{1}{2}\varepsilon\mu^2\delta^{-1}[\eta_t + \varepsilon(q_{x_1}\eta_{x_1} + \gamma^2 q_{x_2}\eta_{x_2})]^2 + \delta^{-3/2}\tilde{\sigma}\left[\mu^2\eta_{x_1x_1}(1+\varepsilon^2\mu^2\gamma^2\eta_{x_2}^2)\right.$$
$$\left.\left. +\mu^2\gamma^2\eta_{x_2x_2}(1+\varepsilon^2\mu^2\eta_{x_1}^2) - 2\varepsilon^2\mu^4\gamma^2\eta_{x_1}\eta_{x_2}\eta_{x_1x_2} - \delta^{3/2}\mu^2(\eta_{x_1x_1} + \gamma^2\eta_{x_2x_2})\right]\right\} \quad (5.17)$$

where

$$\delta = 1 + (\varepsilon\mu)^2(\eta_{x_1}^2 + \gamma^2\eta_{x_2}^2).$$

Equations (5.16) and (5.15) (or its integral form equation (5.16)), are a system of two equations for the two unknowns, η and q.

5.5. *Two dimensional Benney-Luke and KP equations*

In this section, starting from the nonlocal formulation, we derive the Boussinesq, Benney-Luke and KP equations with surface tension. We include surface tension because later we will address the question of fi nding lump type solutions to the water wave equations. We take $|\varepsilon| \ll 1$, $|\mu| \ll 1$, use the expansions

$$\cosh[k\mu(1+\varepsilon\eta)] \sim 1 + \frac{\mu^2}{2}k^2, \quad \sinh[k\mu(1+\varepsilon\eta)] \sim \mu k + \frac{\mu^3}{6}k^3 + \varepsilon\mu\eta k,$$

and we use integration by parts, as well as properties of the Fourier transform

$$k_j \to i\partial x_j, \quad j = 1,2.$$

Then after some algebra we fi nd that to within $O(\varepsilon,\mu^2)$, equations (5.14) and (5.15) yield the following Boussinesq type equations, where for convenience we replace x_1, x_2 by x, y.

$$\left(1 - \frac{\mu^2}{2}\tilde{\Delta}\right)\eta_t + \left(\tilde{\Delta} - \frac{\mu^2}{6}\tilde{\Delta}^2\right)q + \varepsilon\left(\eta_x q_x + \gamma^2\eta_y q_y\right) + \varepsilon\eta\tilde{\Delta}q = 0 \quad (5.18)$$

$$\eta = -q_t - \frac{\varepsilon}{2}\left(q_x^2 + \gamma^2 q_y^2\right) + \tilde{\sigma}\mu^2\tilde{\Delta}\eta, \quad (5.19)$$

where

$$\tilde{\Delta} = \partial_x^2 + \gamma^2 \partial_y^2.$$

Equation (5.19) implies

$$\eta \sim -\left(1 + \tilde{\sigma}\mu^2 \tilde{\Delta}\right) q_t - \frac{\varepsilon}{2}\left(q_x^2 + \gamma^2 q_y^2\right).$$

Substituting this expression in equation (5.18) we find

$$\left[1 + \left(\tilde{\sigma} - \frac{1}{2}\right)\mu^2 \tilde{\Delta}\right] q_{tt} - \left(\tilde{\Delta} - \frac{\mu^2}{6}\tilde{\Delta}^2\right) q + \varepsilon\left(2q_x q_{xt} + 2\gamma^2 q_y q_{yt} + q_t \tilde{\Delta} q\right) = 0. \quad (5.20)$$

Using the leading order approximation, $q_{tt} \sim \tilde{\Delta} q$, equation (5.20) implies the following equation, valid to $O(\varepsilon, \mu^2)$:

$$q_{tt} - \tilde{\Delta} q + (\tilde{\sigma} - \frac{1}{3})\mu^2 \tilde{\Delta}^2 q + \varepsilon(\partial_t |\tilde{\nabla} q|^2 + q_t \tilde{\Delta} q) = 0, \quad (5.21)$$

where $|\tilde{\nabla} q|^2 = \left(q_x^2 + \gamma^2 q_y^2\right)$.

An asymptotically equivalent equation to (5.21) was derived by Benney and Luke [42] in terms of the velocity potential ϕ in the case of zero surface tension ($\tilde{\sigma} = 0$). We note that $q \sim \phi + O(\varepsilon\mu^2)$, which allows us to establish the relationship between the Benney-Luke equation for ϕ and the above equation (5.21) for q.

If we take

$$\gamma = O(\mu), \quad \varepsilon = O(\mu^2),$$

then further simplifications can be made. In this case, (5.21) becomes,

$$q_{tt} - q_{xx} + \left(\tilde{\sigma} - \frac{1}{3}\right)\mu^2 q_{xxxx} - \gamma^2 q_{yy} + \varepsilon\left(2q_x q_{xt} + q_t q_{xx}\right) = 0. \quad (5.22)$$

Letting $\xi = x - t, \quad T = \varepsilon t$, thus $\partial_t = -\partial_\xi + \varepsilon \partial_T, \quad \partial_x = \partial_\xi$.

Then, equation (5.21) simplifies to

$$2\varepsilon q_{T\xi} + \left(\frac{1}{3} - \tilde{\sigma}\right)\mu^2 q_{\xi\xi\xi\xi} + \gamma^2 q_{yy} + 3\varepsilon q_\xi q_{\xi\xi} = 0.$$

Letting $w = q_\xi, \varepsilon = \mu^2 = \gamma^2$, and taking a derivative with respect to ξ, this equation becomes

$$2w_{T\xi} + \left(\frac{1}{3} - \tilde{\sigma}\right) w_{\xi\xi\xi\xi} + w_{yy} + (3ww_\xi)_\xi = 0, \quad (5.23)$$

which is the well known Kadomtsev-Petviashvili (KP) equation [67]. For convenience we put the KP equation (5.23) in standard form by making the following transformations:

$$w = -2\,\mathrm{sgn}(\tilde{\sigma} - \frac{1}{3}) l_1^2 u, \xi \to l_1 x, y \to \frac{1}{\sqrt{3}} y,$$

$$T = -\frac{2\,\mathrm{sgn}(\tilde{\sigma} - \frac{1}{3})\tau}{l_1}, \quad l_1 = |\tilde{\sigma} - \frac{1}{3}|^{\frac{1}{4}};$$

then u satisfies the standard KP equation

$$u_{\tau x} + u_{xxxx} - 3\,\mathrm{sgn}\big(\tilde{\sigma} - \tfrac{1}{3}\big)u_{yy} + 6(uu_x)_x = 0. \tag{5.24}$$

We note that the KdV equation is a reduction of KP by considering the y-independent solutions.

The KP equation is well known to have lump type soliton solutions [cf 43] for $(\tilde{\sigma} - \frac{1}{3}) > 0$. The 1−lump solution is given by the formula

$$u = 16\frac{-4(x' - 2k_R y')^2 + 16k_I^2 y'^2 + \frac{1}{k_I^2}}{\left[4(x' - 2k_R y')^2 + 16k_I^2 y'^2 + \frac{1}{k_I^2}\right]^2}, \tag{5.25}$$

where

$$x' = x - c_x\tau - x_0, \quad y' = y - c_y\tau - y_0, \quad c_x = 12(k_R^2 + k_I^2), \quad c_y = 12k_R;$$

the velocity satisfies the condition

$$c_x > \frac{c_y^2}{12}. \tag{5.26}$$

This has implications regarding the existence of a 1−lump solution for the the Benney-Luke (BL) equation when $(\tilde{\sigma} - \frac{1}{3}) > 0$ (see also the discussion in the computational section). In terms of the original water wave variables, we have

$$x' = \frac{1}{l_1}(\xi + \frac{c_x l_1^2 T}{2}) = \frac{1}{l_1}\left[x - (1 - \frac{l_1^2 \varepsilon c_x}{2})t\right], \tag{5.27a}$$

$$y' = \sqrt{3}(x_2 + \frac{c_y l_1 T}{2\sqrt{3}}) = \sqrt{3}(y + \frac{l_1 \varepsilon c_y t}{2\sqrt{3}}). \tag{5.27b}$$

Calling $v_x = 1 - \varepsilon\tilde{c_x}$ and $v_y = \varepsilon\tilde{c_y}$, it follows that if $\tilde{c_x}$ and $\tilde{c_y}$ are defined by $\tilde{c_x} = c_x l_1^2/2$ and $\tilde{c_y} = c_y l_1/(2\sqrt{3})$, then

$$\tilde{c_x} > \frac{\tilde{c_y}^2}{2},$$

which is consistent with the previous condition of equation (5.26).

Comparing the derivation of the KP equation (in terms of u) with the derivation of the BL equation (in terms of q) it follows that that q and u are related by

$$q_x = w = -2\,\mathrm{sgn}\big(\tilde{\sigma} - \tfrac{1}{3}\big)\,\big|\tilde{\sigma} - \tfrac{1}{3}\big|^{\frac{1}{2}}\,u. \tag{5.28}$$

Thus traveling waves for KP and BL equations are directly related, provided that their velocities are related via $\tilde{c_x} = c_x l_1^2/2$ and $\tilde{c_y} = c_y l_1/(2\sqrt{3})$. An explicit lump formula for q is obtained by integrating equations (5.25) and (5.28) with respect to x to find

$$q = -32(\tilde{\sigma} - \frac{1}{3})^{\frac{1}{2}} \left[\frac{(x' - 2k_R y')}{4(x' - 2k_R y')^2 + 16k_I^2 y'^2 + \frac{1}{k_I^2}} \right]. \tag{5.29}$$

The above equations imply that as $r \to \infty$, $u \sim O(1/r^2), q \sim O(1/r)$, where $r^2 = x'^2 + y'^2$.

5.6. *Computational Studies — 2+1 dimensional lumps*

In this section we discuss numerical solutions of lump type traveling wave solutions to the Benney-Luke and fully nonlinear $2+1$ dimensional water wave equations. The numerical solutions are obtained from the fixed point SPRZ method described earlier. A detailed discussion can be found in Ablowitz et al. [41].

When considering a traveling solution of the BL equation (5.21) we assume

$$q(x,y,t) = q(x - v_x t, y - \gamma^2 v_y t) = q(x', y').$$

where where the BL velocities (v_x, v_y) are given by,

$$v_x = 1 - \varepsilon\tilde{c_x}, \quad \gamma^2 v_y = \gamma^2 \tilde{c_y} \tag{5.30}$$

and we specify $\tilde{c_x}, \tilde{c_y}$. It should be noted that the BL velocities are connected to KP velocities (c_x, c_y), by

$$\tilde{c_x} = \frac{c_x}{2}(\tilde{\sigma} - \tfrac{1}{3})^{\frac{1}{2}}, \tilde{c_y} = \frac{c_y}{2\sqrt{3}}(\tilde{\sigma} - \tfrac{1}{3})^{\frac{1}{4}}. \tag{5.31}$$

The same numerical method used to find the traveling wave solution for the BL equation also works for KP. The KP equation also provides a check of the numerical method used herein since KP admits an explicit lump solution.

In figure 6 we present a typical example of a wave profile associated with the KP and BL equation; plotted are the $y = 0, x = 0$ profiles respectively, when $c_x = 3, c_y = 0$. The exact solution is obtained from equation (5.25), and the numerical results are found using the SPRZ method introduced in the previous section. The numerical solution associated with KP cannot be distinguished from the theoretical solution in the graph; the numerical scheme correctly reproduces the known KP lump -soliton solution. We have plotted the wave profiles at $y = 0$ and $x = 0$ respectively corresponding to the associated "KP" function,

$$u = -\frac{q_x}{2(\tilde{\sigma} - \frac{1}{3})^{\frac{1}{2}}}, \tag{5.32}$$

The figure shows that KP is a good approximation to BL in this range of parameters.

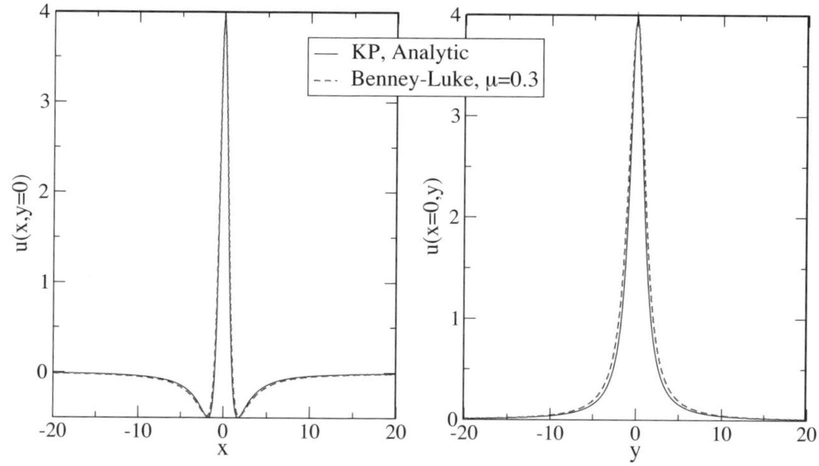

Figure 6. Wave profile at $y = 0$ for the KP (5.24) and BL equations (5.21) with $\tilde{\sigma} = 2/3$, $c_x = 3$, $c_y = 0$.

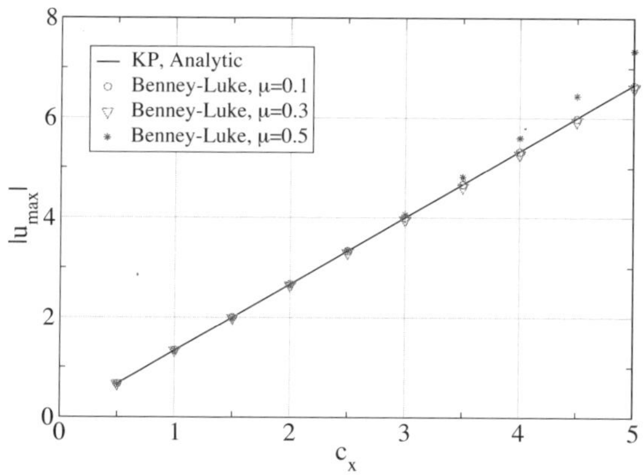

Figure 7. $u(0,0) = u_{max}$ vs. c_x for various values of μ. This figure also shows that the KP theory is a good approximation to the BL equation in this range of parameters.

Figure 7 shows a nearly linear relationship between u and c_x (note in figure 7, $v_x = 1 - \varepsilon\tilde{c_x}$, $v_y = 0$; see equation (5.31) for the relationship between c_x and $\tilde{c_x}$). Comparison with the KP theory of equation (5.25) is given by the solid line. Even at $\mu = 0.5$, the KP solution provides a good approximation to the BL lump solution though we see that there is some deviation at large values of c_x.

Next, we look for two-dimensional traveling wave lumps associated with the full two plus one dimensional water wave equations. As with the BL equation, we consider traveling wave solutions of the form $\eta(x', y')$, $q(x', y')$ where $x' - v_x t$, $y' = y - \gamma^2 v_y t$ with v_x, v_y satisfying the relation (5.30).

In figures 8 and 9 solutions of the full water wave equations are depicted. Numerical solutions are obtained employing the method described earlier; theoretical values are calculated from equation (5.25). In Fig. 8, the wave profiles at $y = 0$ and $x = 0$ respectively (note we drop the $'$), corresponding to the associated "KP" function u obtained from equation (5.32), are also given. The dotted lines lie on top of the theoretical result hence are difficult to discern.

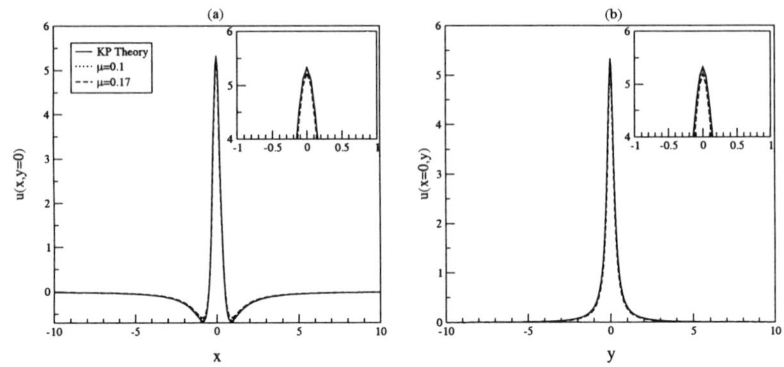

Figure 8. Wave profile at a) $y = 0$ and b) $x = 0$ for the full water wave equations (5.16–5.17) with $\tilde{\sigma} = \frac{2}{3}$. The function u is plotted using equation (5.32) and the velocities are $v_x = 1 - \varepsilon\tilde{c_x}, v_y = 0$ with $c_x = 4.0$ using the relationship from equation (5.31). This figure demonstrates that Benney-Luke/KP equations are good approximations to the full water wave equations. The inset shows an enlarged view of the peak of the wave, highlighting the difference in maximum amplitudes.

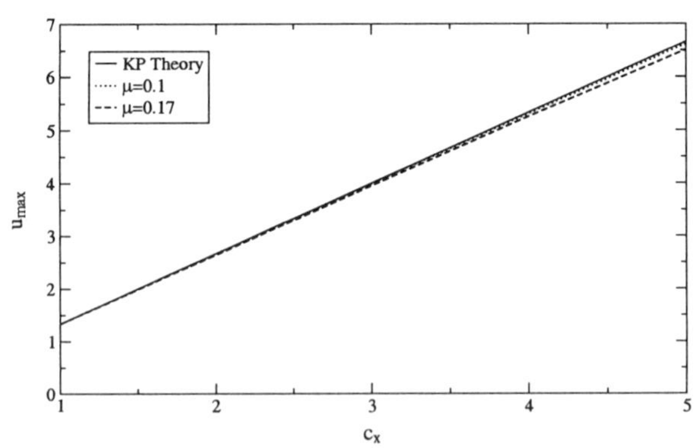

Figure 9. $u(0,0) = u_{max}$ vs. c_x the full water wave equations Eqs. (5.16–5.17) for various values of μ using the same parameters as Fig. 8. The same function u as in Fig. 8 is plotted here. This figure also shows that the Benney-Luke/KP equations are good approximations to full water waves for $\mu = 0.1$.

To summarize, a coupled system of nonlocal nonlinear equations on a fixed domain is derived which governs the classical free boundary equations of water waves in $2+1$ dimensions with constant depth. In the shallow-water/long-wave limits, these equations asymptotically reduce to the Benney-Luke and Kadomtsev-Petviashvili (KP) equations. The SPRZ technique is used to compute solitary-lump solutions for the Benney-Luke/KP and the nonlocal $2+1$ dimensional water wave equations. The Fourier nature of the equations is very useful in this computational evaluation.

Acknowledgments: This work was partially supported by NSF grants DMS 0303756, DMS0505352.

References

1. G. P. Agrawal. *Nonlinear Fiber Optics*. Third Edition. Academic Press, New York (2001).
2. R. W. Boyd. *Nonlinear Optics*. Second Edition. Elsevier, London (2002).
3. H. Lamb. *Hydrodynamics*. Dover Publications, New York (1932).
4. J. J. Stoker. *Water Waves: The Mathematical Theory with Applications*. John Wiley and Sons, New York (1958).
5. J. S. Russell. Report on waves. In *Report of the 14th meeting of the British Association*, pages 311–390. John Murray, London (1845).
6. J. Boussinesq. *Mémoires présentés par divers savants à l'Académie des Sciences*, **–**, 1–680 (1877).
7. D.J. Korteweg and G. de Vries. *Phil. Mag.*, **39**, 422–443 (1895).
8. A. Hasegawa and F. D. Tappert. *Applied Physics Letters*, **23**, 142–144 (1973).
9. A. Hasegawa and F. D. Tappert. *Applied Physics Letters*, **23**, 171–172 (1973).
10. L. F. Mollenauer, R. H. Stolen, and J. P. Gordon. *Physical Review Letters*, **45**, 1095–1098 (1980).
11. A. Hasegawa and Y. Kodama. *Solitons in Optical Communciations*. Oxford University Press, Oxford (1995).
12. L. F. Molleneauer and J. P. Gordon. *Solitons in Optical Fibers: Fundamental and Applications to Telecommunications*. Academic Press, London (2006).
13. C. Lin, H. Kogelnik, and L. G. Cohen. *Optics Letters*, **5**, 476–478 (1980).
14. J. H. B. Nijhof, N. J. Doran, W. Forysiak, and F. M. Knox. *Electronics Letters*, **33**, 1726–7 (1997).
15. P. V. Mamyshev and L. F. Mollenauer. *Optics Letters*, **21**, 396–398 (1996).
16. M. J. Ablowitz, G. Biondini, S. Chakravarty, and R. L. Horne. *Journal of the Optical Society of America B*, **20**, 831–845 (2003).
17. M. J. Ablowitz, A. Docherty, and T. Hirooka. *Optics Letters*, **28**, 1191–1193 (2003).
18. M. J. Ablowitz and G. Biondini. *Optics Letters*, **23**, 1668–70 (1998).
19. M. J. Ablowitz, G. Biondini, and E. Olson. On the evolution and interaction of dispersion-managed solitons. In *Massive WDM and TDM Soliton Transmission Systems*, edited by Akira Hasegawa, pages 362–7, Kyoto, Japan (2000). Kluwer Academic Publishers.
20. I. R. Gabitov and S. K. Turitsyn. *Optics Letters*, **21**, 327–329 (1996).

21. V. Zharnitsky, E. Grenier, S. K. Turitsyn, C. K. Jones, and J. S. Hesthaven. *Physical Review E*, **62**, 7358–64 (2000).
22. M. J. Ablowitz, T. Hirooka, and G. Biondini. *Optics Letters*, **26**, 459–461 (2001).
23. M. J. Ablowitz and T. Hirooka. *Journal of the Optical Society of America B*, **19**, 425–439 (2002).
24. V. I. Petviashvili. *Soviet Journal of Plasma Physics*, **2**, 254–263 (1976).
25. D. E. Pelinovsky and Y. A. Stepanyants. *SIAM Journal on Numerical Analysis*, **42**, 1110–1127 (2004).
26. M. J. Ablowitz and Z. H. Musslimani. *Optics Letters*, **30**, 2140–2142 (2005).
27. M. J. Ablowitz, G. Biondini, and S. Blair. *Physics Letters A*, **236**, 520–524 (1997).
28. M. J. Ablowitz, G. Biondini, and S. Blair. *Physical Review E*, **63**, 046605 (2001).
29. D. J. Benney and G. J. Roskes. *Studies in Applied Mathematics*, **48**, 377–385 (1969).
30. V. D. Djordjevic and L. G. Redekopp. *Journal of Fluid Dynamics*, **79**, 703–714 (1977).
31. A. Davey and Stewarts.K. *Proceedings Of The Royal Society Of London Series A*, **338** , 101–110 (1974).
32. M. S. Longuet-Higgens and E. D. Cokelet. *Proceedings Of The Royal Society Of London Series A*, **1**, 350 (1976).
33. B. Fornberg. *SIAM Journal on Scientific and Statistical Computing*, **1**, 386–400 (1980).
34. J. W. Dold. *Journal Of Computational Physics*, **103**, 90–115 (1992).
35. V. E. Zakharov, A. I. Dyachenko, and O. A. Vasilyev. *European Journal Of Mechanics B*, **21**, 283–291 (2002).
36. V. E. Zakharov. *Journal of Applied Mechanics and Technical Physics*, **2**, 190–194 (1968).
37. W. Craig and C. Sulem. *Journal Of Computational Physics*, **108**, 73–83 (1993).
38. W. Craig and M. D. Groves. *Wave Motion*, **19**, 367–389 (1994).
39. W. Craig and D. P. Nicholls. *SIAM Journal on Mathematical Analysis*, **32**, 323–359 (2000).
40. W. J. D. Bateman, C. Swan, and P. H. Taylor. *Journal of Computational Physics*, **174**, 277–305 (2001).
41. M. J. Ablowitz, A. S. Fokas, and Z. H. Musslimani. On a new nonlocal formulation of water waves. Accepted for publication (2006).
42. D. J. Benney and J. C. Luke. *Journal of Mathematics and Physics*, **43**, 455 (1964).
43. M. J. Ablowitz and P. A. Clarkson. *Solitons, Nonlinear Evolution Equations and Inverse Scattering.* Cambridge University Press, Cambridge, UK. (1991).
44. M. J. Ablowitz and Z. H. Musslimani. *Physical Review E*, **67**, 025601 (2003).
45. M. J. Ablowitz, B. Ilan, and S. T. Cundiff. *Optics Letters*, **29**, 1808–1810 (2004).
46. Q. Quraishi, S. T. Cundiff, B. Ilan, and M. J. Ablowitz. *Physical Review Letters*, **94**, 243904 (2005).
47. M. Ablowitz, I. Bakirtas, and B. Ilan. *Physica D*, **207**, 230–253 (2005).
48. P. L. Kelley. *Physical Review Letters*, **15**, 1005 (1965).
49. S. Vlasov, V. Petrishchev, and V. Talanov. *Radiophysics and Quantum Electronics*, **14** (1971).

50. M. I. Weinstein. *Communications in Mathematical Physics*, **87**, 567 (1983).
51. R. Y. Chiao, E. Garmire, and C. H. Townes. *Physical Review Letters*, **13**, 479 (1964).
52. G. C. Papanicolaou, D. McLaughlin, and M. Weinstein. *Lecture Notes in Numerical and Applied Analysis*, **5**, 253 (1982).
53. C. Sulem and P. L. Sulem. *The Nonlinear Schrödinger Equation*. Springer-Verlag, New York (1999).
54. F. Merle and P. Raphael. *Inventiones Mathematicae*, **156**, 565–672 (2004).
55. K. D. Moll, A. L. Gaeta, and G. Fibich. *Physical Review Letters*, **90**, 203902 (2003).
56. M. J. Ablowitz and R. Haberman. *Physical Review Letters*, **35**, 1185–1188 (1975).
57. M. J. Ablowitz and H. Segur. *Journal Of Fluid Mechanics*, **92**, 691–715 (1979).
58. A. S. Fokas and M. J. Ablowitz. *Physical Review Letters*, **51**, 7–10 (1983).
59. M. J. Ablowitz and H. Segur. *Solitons and the Inverse Scattering Transform*. SIAM Publications, Philadelphia, Philadelpia (1981).
60. J. M. Ghidaglia and J. C. Saut. *Nonlinearity*, **3**, 475–506 (1990).
61. R. Cipolatti. *Communications in Partial Differential Equations*, **17**, 967–988 (1992).
62. G. C. Papanicolaou, C. Sulem, P. L. Sulem, and X. P. Wang. *Physica D*, **72**, 61–86 (1994).
63. F. Merle and Y. Tsutsumi. *Journal of Differential Equations*, **84**, 205–214 (1990).
64. M. Mitchell, M. Segev, T. H. Coskun, and D. N. Christodoulides. *Physical Review Letters*, **79**, 4990–4993 (1997).
65. M. J. Ablowitz and Z. Musslimani. *Physica D*, **184**, 276–303 (2003).
66. M. A. Hoefer, M. J. Ablowitz, E. A. Coddington, I. Cornell, P. Engels, and V. Schweikhard. On dispersive and classical shock waves in bose-einstein condensates and gas dynamics. Accepted for publication (2006).
67. B. B. Kadomtsev and V. I. Petviashvili. *Doklady Akademii Nauk SSSR*, **192**, 753–756 (1970).

RESONANCE PROBLEMS IN PHOTONICS

M. I. WEINSTEIN
Department of Applied Physics and Applied Mathematics
Columbia University
New York, NY 10027
Email: miw2103@columbia.edu

1. Introduction and outline

An important area of photonics concerns the propagation and control of light in non-homogeneous or nonlinear media. This is an area of great importance due to the range of physical effects to be exploited in the design of devices for communication and computing, as well as in the study of fundamental phenomena. Progress has involved a rich interplay of experiment, modeling, computation and analysis, which have impact beyond the specific motivating questions. Indeed, the models and analysis that we discuss are closely related those arising in the study of Bose-Einstein condensation in the regime governed by Gross-Pitaevskii type equations.

The propagation of light in a medium is governed by Maxwell's equations with appropriate linear and nonlinear constitutive relations. We consider a setting where there is a separation of scales in the electric field; the carrier wavelength is short compared to a wave envelope width; see Figure 1. An approximate description of the field evolution, in terms of its *envelope* is therefore very natural and effective. In this article we discuss various resonance phenomena leading to and playing a role in the dynamics of such envelope equations. Our discussion is based, in part, on joint papers with R.H. Goodman, P.J. Holmes, R.E. Slusher and A. Soffer; see the references.

We begin with the modeling of optical pulses traveling through a nonlinear and periodic medium. At *Bragg resonance*, when the carrier wavelength and medium periodicity satisfy a resonance relation, the wave envelope evolves according to a system of dispersive nonlinear partial differential equations (PDEs), the nonlinear coupled mode equations (NLCME). NLCME has *gap soliton* solutions. These are localized states with frequencies in a photonic band gap. Theory predicts that these can travel at any speed less than the speed of light. It is of fundamental and practical interest to understand whether one can "trap" such soliton pulses. This would have potential application to optical buffers and computing [20].

We explore strategies for soliton capture via the insertion of suitably designed localized defects. The dynamics are governed by a variant of NLCME, but now with

occurs when the carrier wavelength, λ is twice the medium periodicity, d:

$$\lambda_B = 2d, \quad k_B = \frac{2\pi}{\lambda_B} = \frac{\pi}{d}. \tag{2.4}$$

For small ε the solution can be represented, on non-trivial time-scales, by (2.3), where $E_\pm$ are no longer constants, but rather evolve slowly according to coupled mode equations, a system of PDEs , which tracks the coupled evolution of backward and forward waves. As we shall see below, propagation in a periodic structure is *dispersive*; waves of different wavelengths travel at different speeds.

We now introduce nonlinearity through the refractive index (2.2), $n_2 > 0$, and assume field amplitudes, chosen to balance the linear index modulations. This establishes a balance between dispersive and nonlinear effects. We encode the multiple scale nature of the problem in terms of a small parameter ε, by the relations:

$$\text{Pulse width} \sim \varepsilon^{-1}, \quad \text{Pulse amplitude} \sim \varepsilon^{1/2}, \quad \text{Carrier wavelength} \sim 1;$$

see Figure 2. The electric field then has the form

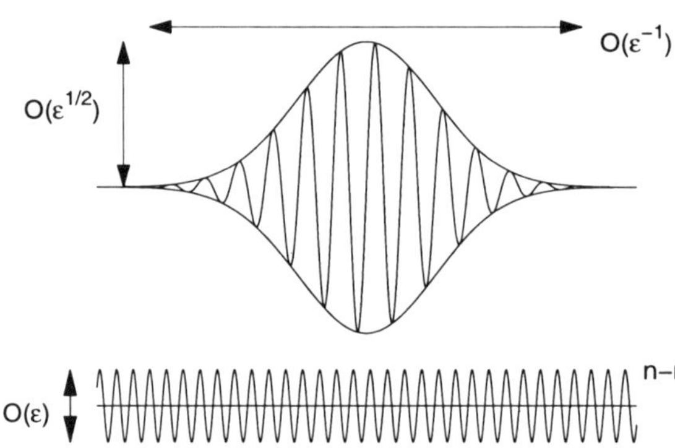

Figure 2. Scalings of pulse width, amplitude and carrier wavelength.

$$\begin{aligned} E^\varepsilon &\sim \varepsilon^{\frac{1}{2}} \left(E_+(Z,T)e^{i(k_B z - \omega_B t)} + E_-(Z,T)e^{-i(k_B z + \omega_B t)} \right) \quad + \quad c.c. \\ Z &= \varepsilon z, \ T = \varepsilon t, \end{aligned} \tag{2.5}$$

where the amplitudes $E_+(Z,T)$ and $E_-(Z,T)$ satisfy the *Nonlinear Coupled Mode Equations* (NLCME)

$$\begin{aligned} & i(\partial_T + \partial_Z)E_+ + \kappa E_- + g(|E_+|^2 + 2|E_-|^2)E_+ = 0 \\ & i(\partial_T - \partial_Z)E_- + \kappa E_+ + g(|E_-|^2 + 2|E_+|^2)E_- = 0, \quad g > 0. \end{aligned} \tag{2.6}$$

See, for example, Refs. 17 for an overview. A result on the validity of NLCME, as an approximation to solutions of a nonlinear wave system at Bragg resonance, is the following [17]:

Theorem 2.1. *Solutions $E_\pm(Z,T)$ of the nonlinear coupled mode equations (NLCME) yield $H^1(\mathbf{R})$ accurate approximations to solutions of the Anharmonic Maxwell-Lorentz model on a time scale of physical interest. Specifically,*

$$E^\varepsilon(z,t) \; - \; \Re\, \varepsilon^{\frac{1}{2}} \left(E_+(\varepsilon z, \varepsilon t)e^{i(k_B z - \omega_B t)} + E_-(\varepsilon z, \varepsilon t)e^{-i(k_B z + \omega_B t)}\right)$$

is $\mathcal{O}(\varepsilon)$ in $H^1(\mathbf{R})$ for $t \sim \mathcal{O}(\varepsilon^{-1})$,

3. Coherent structures

Note that the linearization of (2.6) about the zero state leads to a linear *dispersive* equation:

$$i(\partial_T + \partial_Z)E_+ + \kappa E_- = 0, \qquad i(\partial_T - \partial_Z)E_- + \kappa E_+ = 0. \tag{3.1}$$

Indeed, (3.1) has plane waves of the form: $\vec{v}e^{-i\Omega T}$ with resulting dispersion relation, $\Omega = \Omega(Q)$ given by

$$\Omega^2 = Q^2 + \kappa^2.$$

Note the gap in the spectrum of admissible plane waves; there are no plane waves with frequencies in the interval $[-\kappa, \kappa]$. This corresponds to the fundamental (lowest energy) gap in the spectrum of the Helmholtz wave equation, associated with (2.1) for $n_2 = 0$.

Solitary wave solutions of (2.6) can be sought in the form $(E_-(Z;\omega), E_+(Z;\omega))e^{-i\omega t}$. Explicit solutions were given in Ref. 9, 1, by a generalization of the soliton-form for the massive Thirring (integrable) equation . For each ω in the interval (photonic band gap) $(-\kappa, \kappa)$, there is a solitary wave of finite L^2 norm (optical power). Therefore, these solitary waves are called *gap solitons.* A bifurcation diagram for gap solitons is displayed in Figure 3. In fact, a family of gap solitons can be found, which travel at any speed $0 \le v < c$.

Experiments by Slusher *et. al.* (1997) have explored nonlinear propagation of light in periodic structures at *Bragg resonance* in fiber gratings. They observe gap soliton propagation with speeds as slow as 50% of the speed of light [14].

Can a gap solitons propagating in a periodic structure be captured by insertion of appropriately engineered defects?

The approach to this question, taken in Ref. 19, is discussed in the following section and leads to some basic questions in the nonlinear scattering of coherent structures.

4. Controlling nonlinear pulses

A simple model of a nonlinear periodic structure with a localized defect is given by the generalization of (2.2)

$$n^2 \; = \; n_0^2(\varepsilon z) \; + \; \varepsilon\kappa(\varepsilon z)\cos\left(\frac{2\pi}{d}z + \Phi(\varepsilon z)\right) \; + \; n_2 E^2 \tag{4.1}$$

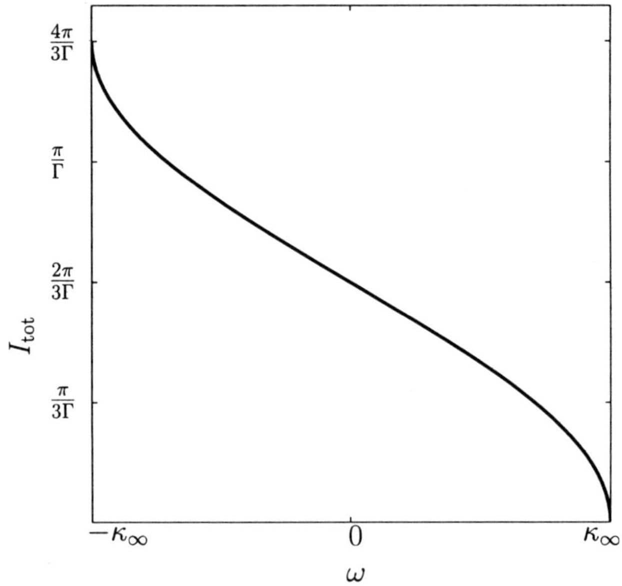

Figure 3. $I = \int |E_+|^2 + |E_-|^2 \, dZ$ vs. ω for $\omega \in (-\kappa, \kappa)$.

Here, $n_0(Z), \kappa(Z)$ and $\Phi(Z)$ are now slowly varying parametric functions, which approach constant limits away from some some compact set. Schematic plots of background linear refractive indices are shown in Figure 4. Now, in a manner

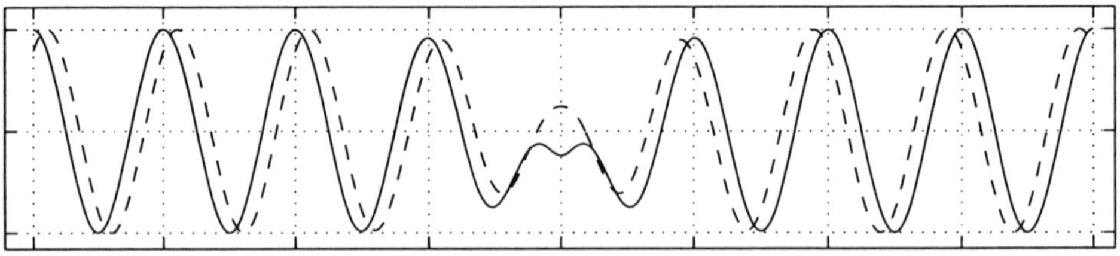

Figure 4. Linear refractive index of a periodic medium with a localized defect for two sets of parametric functions $n_0(\varepsilon z), \kappa(\varepsilon z)$ and $\Phi(\varepsilon z)$

analogous to the derivation of NLCME (2.6), we obtain the following generalization governing propagation in periodic media with defects [19]:

$$\begin{aligned} i(\partial_T + \partial_Z)E_+ + \kappa(Z)E_- + W(Z)E_+ + g(|E_+|^2 + 2|E_-|^2)E_+ = 0 \\ i(\partial_T - \partial_Z)E_- + \kappa(Z)E_+ + W(Z)E_- + g(|E_-|^2 + 2|E_+|^2)E_- = 0 \end{aligned} \quad (4.2)$$

The parametric function $W(Z)$ can be computed from the functions $\kappa(\cdot), \Phi(\cdot)$ and $n_0(\cdot)$, which define the linear refractive index [19].

4.1. *Nonlinear scattering*

We have numerically investigated the interaction of incoming gap solitons, of different amplitudes and phases, with a variety of defects. Some representative results [19] of such simulations appear in Figures 5 and 6; see also Ref. 18. Simulations show

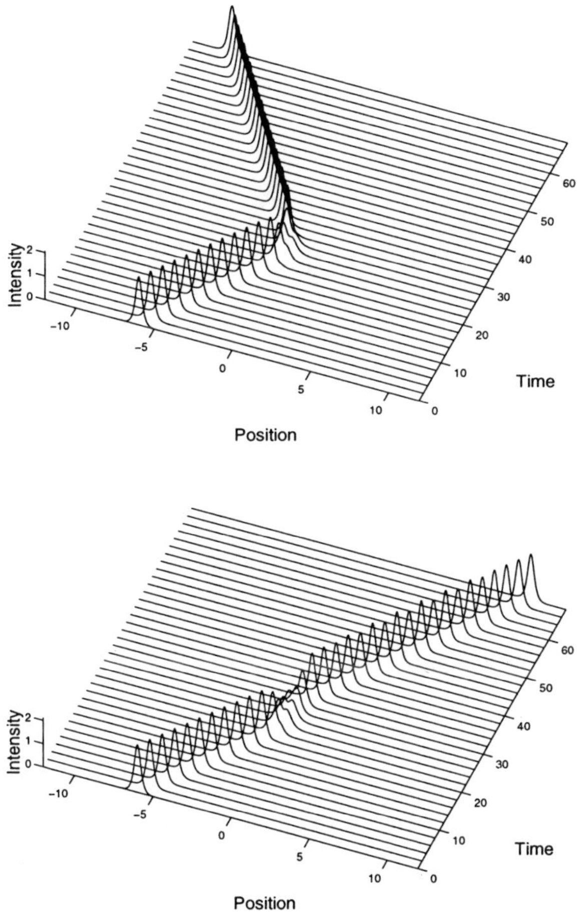

Figure 5. Reflection and transmission of gap soliton

a soliton incident on the defect and the resulting redistibution of its energy into (a) a reflected soliton, (b) localized states within the defect region, (c) a transmitted soliton and (d) outgoing radiation.

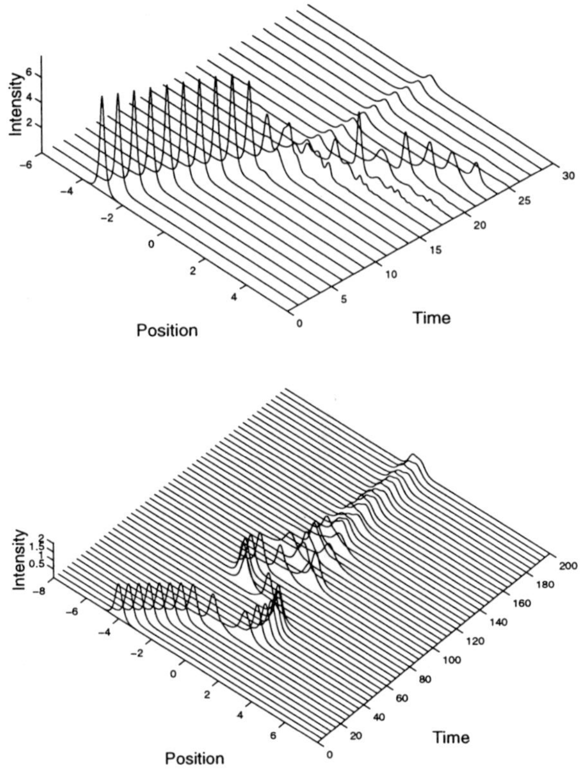

Figure 6. Partial capture and capture of gap soliton

We next identify these localized (pinned) states within the defect region as being *nonlinear defect modes* related to the background linear refractive index.

4.2. *Nonlinear defect modes*

Nonlinear defect modes are standing wave solutions of (4.2), which are spatially localized ("pinned") at the defect. Setting $(E_+(Z,T), E_-(Z,T) = (\mathcal{E}_+(Z), \mathcal{E}_Z(Z))e^{-i\omega T}$, we have

$$(\omega + i\partial_Z)\mathcal{E}_+ + \kappa(Z)E_- + W(Z)\mathcal{E}_+ + g(|\mathcal{E}_+|^2 + 2|\mathcal{E}_-|^2)\mathcal{E}_+ = 0$$
$$(\omega - i\partial_Z)\mathcal{E}_- + \kappa(Z)E_+ + W(Z)\mathcal{E}_- + g(|\mathcal{E}_-|^2 + 2|\mathcal{E}_+|^2))\mathcal{E}_- = 0 \qquad (4.3)$$

Consider first the linearization of (4.2) about the zero state. We have

$$i(\partial_T + \partial_Z)E_+ + \kappa(Z)E_- + W(Z)E_+ = 0$$
$$i(\partial_T - \partial_Z)E_- + \kappa(Z)E_+ + W(Z)E_- = 0 \qquad (4.4)$$

Linear defect modes are solutions of (4.4) of the form $E_{\pm} = e^{-i\Omega T}F_{\pm}(Z;\Omega)$ with $F \in L^2$, or equivalently are L^2 eigenstates of the self-adjoint linear operator,

$$i\sigma_3\partial_Z + \kappa(Z)\sigma_2 + W(Z),$$

where σ_2 and σ_3 are standard Pauli matrices. Thus,

$$\begin{aligned}(\Omega + i\partial_Z)F_+ + \kappa(Z)F_- + W(Z)F_+ &= 0\\(\Omega - i\partial_Z)F_- + \kappa(Z)F_+ + W(Z)F_- &= 0\end{aligned} \qquad (4.5)$$

Remark 4.1. Illustrated in Figure 4 are two different linear periodic structures with defects, (4.1) with $n_2 = 0$, for which the (envelope) spectral problem (4.5) is identical.

In analogy with a result for the nonlinear Schrödinger / Gross-Pitaevskii equation (see section 5) [32], we have [19]:

Theorem 4.1. *Nonlinear defect modes of (4.2) bifurcate from the zero state at the linear discrete eigenvalues of the eigenvalue problem (4.5). These are standing wave states which are localized or "pinned" at the defect site.*

Figure 7 shows two branches of states. The dark curve corresponds to gap solitons for translation invariant equation (2.6); see also Figure 3. These bifurcate from the zero state at the edge of the continuous spectrum. The lighter curve corresponds to a family of nonlinear defect states bifurcating from the zero state at the eigenvalue, $\omega = -1$, of the linear eigenvalue problem (4.5). Such states have recently been considered as well for a two-dimensional variant of (4.2); see Ref. 13. Note that if the linear problem (4.5) has multiple defect states, then there are multiple branches bifurcating from the linear eigenvalues [19,21]. In section 5 we consider, for the NLS-GP equation, the implications for the detailed nonlinear dynamics of multiple branches of defect states.

The problem of capturing a gap soliton by a defect can be seen in terms of the transfer of energy of an incoming soliton to nonlinear defect states. For low velocity incoming solitons, we find that an incoming soliton transfers its energy to a nonlinear defect state (is trapped) if there is an energetically accessible (lower L^2 norm) defect state of approximately the same frequency with which it can resonate [19,18]; see Figure 7. Figures 8 and 9 show the location of the soliton center of mass, z_{cm}, as a function of time in the non-resonant and resonant cases. In the non-resonant case, the soliton has been reflected, while in the resonant case, much of its energy has been trapped. The lower plots are of mode projections and indicate the degree of energy exchange between soliton and defect mode.

A rigorous theory of soliton capture is an open problem. In the recent papers, Refs. 25, 26, fast soliton scattering from a delta-function defect is studied analytically and numerically.

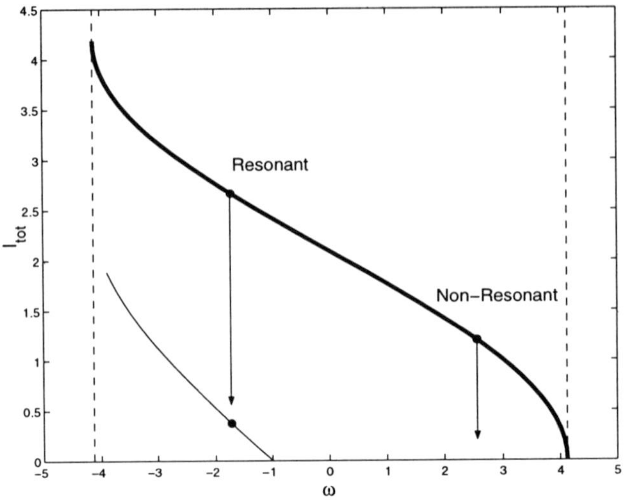

Figure 7. Bifurcation curves and resonant energy transfer

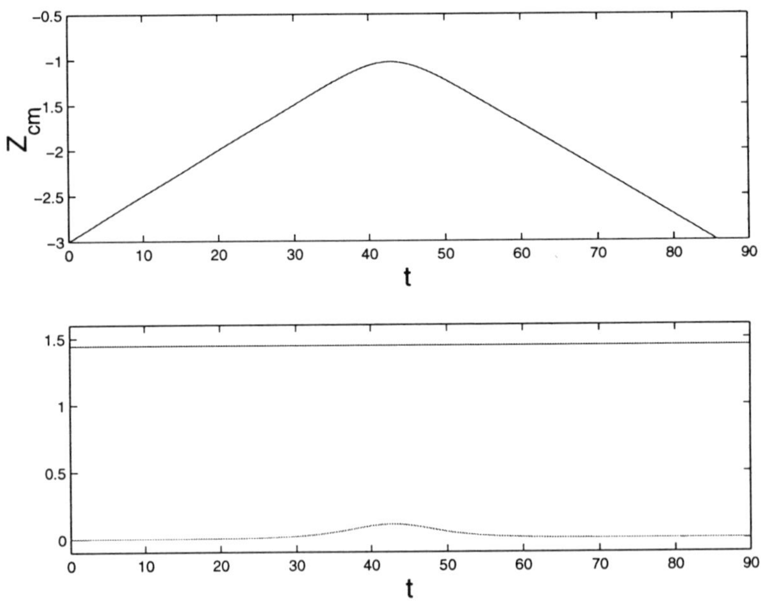

Figure 8. Nonresonance - minimal energy transfer to defect mode

5. Large time evolution of a captured soliton - ground state selection for NLS / G-P

Suppose a soliton is trapped by a defect, say as in Figure 6. The defect may have multiple linear bound states and therefore multiple bifurcating nonlinear bound

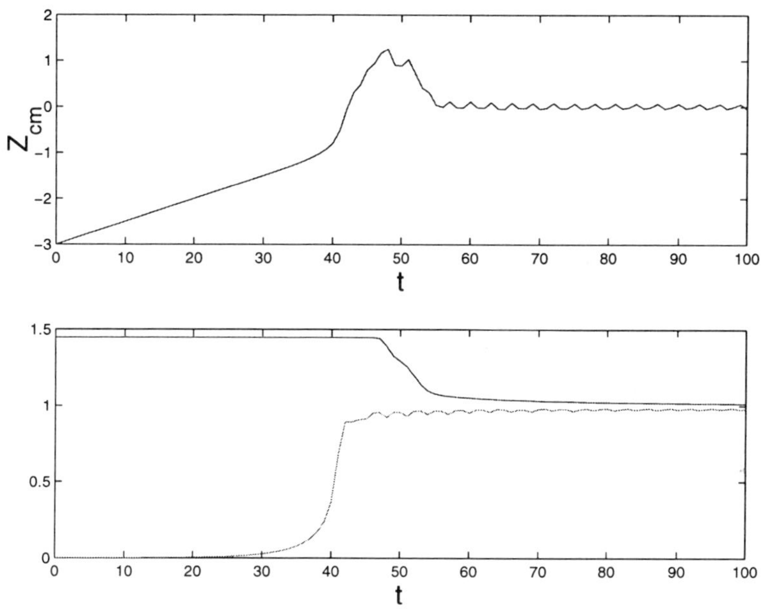

Figure 9. Resonance: Strong energy transfer to defect mode

state families. These compete for the soliton's energy. In this section we consider the question of how the energy distributes among the available modes in the large time limit.

This is a question of general interest and we consider it in the context of the nonlinear Schrödinger / Gross-Pitaevskii (NLS-GP) equation:

$$\begin{aligned} i\partial_t\Phi &= (\ -\Delta \ + \ V(x)\ )\,\Phi \ + \ g|\Phi|^2\Phi \\ &= H\Phi \ + \ g|\Phi|^2\Phi \qquad (5.1) \\ \Phi(x,0) &= \Phi_0(x),\ x \in \mathbb{R}^3, \ \ \text{small norm and localized.} \end{aligned}$$

Here, $V(x)$ is a potential which decays rapidly as $|x| \to \infty$.

NLS-GP can be derived in the context of nonlinear optics [3,42] and in the study of the macroscopic behavior of a large ensemble of quantum particles (Bosons), *e.g.* Bose-Einstein condensates (BEC); see, for example, Refs. 15, 2. In (5.1), g is a constant, corresponding to, in nonlinear optics, the negative of the Kerr nonlinear coefficient and, for Bose-Einstein condensation, the scattering length of the inter-particle quantum potential .

Assume that H has two bound states:

$$H\psi_{j*} \ = \ E_{j*}\psi_{j*}, \ \ \|\psi_{j*}\|_2 = 1, \ \ j = 0, 1$$

First consider the linear case, where $g = 0$. If Φ_0 is sufficiently localized, then the solution decomposes into a spatially localized, time quasi-periodic part and a part which decays to zero (in L^∞) [27,47] due to dispersion, as $t \to \pm\infty$:

$$\Phi(x,t) \;=\; \underbrace{c_0 e^{-iE_{0*}t}\psi_{0*}(x) \;+\; c_1 e^{-iE_{1*}t}\psi_{1*}(x)}_{\text{quasi-periodic}} \;+\; \mathcal{O}(t^{-\frac{3}{2}}).$$

How does the solution of NLS/G-P ($g \neq 0$) resolve as $t \to \infty$?

To study this question, we begin by introducing the *nonlinear bound states* or *defect modes* of the NLS / Gross Pitaevskii Equation. These are solutions of (5.1) of the form $\Psi(x;E)e^{-iEt}$, where

$$H\Psi \;+\; g|\Psi|^2\Psi \;=\; E\Psi$$

The following result, analogous to Theorem 4.1, shows that there are families of nonlinear bound states, which bifurcate from the discrete eigenstates of H [32].

Theorem 5.1. *There exist nonlinear bound states bifurcating from the zero state at the linear eigenvalues, E_{j*}.*

$$\begin{aligned} \Psi_{\alpha_j} &= \alpha_j\left(\psi_{j*} \;+\; \mathcal{O}(|\alpha_{j*}|^2)\right), \;\; \alpha_j \in \mathbf{C} \\ E_j &= E_{j*} \;+\; \mathcal{O}(|\alpha_{j*}|^2), \;\; |\alpha_j| \to 0 \end{aligned} \tag{5.2}$$

How do the nonlinear bound states Ψ_{α_0} and Ψ_{α_1} participate in the dynamics on small, intermediate and infinite time scales?

The following theorem addresses the above questions [40,41].

Theorem 5.2. *For generic small initial data, the solution of the nonlinear Schrödinger / Gross-Pitaevskii equation (NLS-GP) (5.1) evolves toward a nonlinear ground state as $t \to \pm\infty$.*

More precisely, assume

- *(H1) H has 2 bound states ($\Longrightarrow$ 2 families of nonlinear defect modes)*
- *(H2) Φ_0 is smooth and spatially localized and small (weakly nonlinear regime)*
- *(H3) $\Gamma \;\equiv\; \pi\left|\mathcal{F}_{H_c}[\psi_{0*}\psi_{1*}^2](\omega_{res})\right|^2 \;>\; 0$, where $\omega_{res} = 2E_{1*} - E_{0*}$; (nonlinear variant of Fermi Golden Rule). Here, $\mathcal{F}_{H_c}[f](\omega_{res})$ denotes the projection of f onto the generalized plane-wave eigenfunction of H (generalized Fourier transform) at frequency ω_{res}.*

Then,

(1) As $t \to \pm\infty$,

$$\Phi(t) \;=\; e^{i\omega_j^\pm(t)}\Psi_{\alpha_j^\pm}(x) \;+\; \mathcal{O}(t^{-\frac{1}{2}})$$

where either $j = 0$ (nonlinear ground state) or $j = 1$ (nonlinear excited state).

(2) Generically, $j = 0$. $\omega_0^\pm(t) = E_0^\pm t + \mathcal{O}(\log t)$.

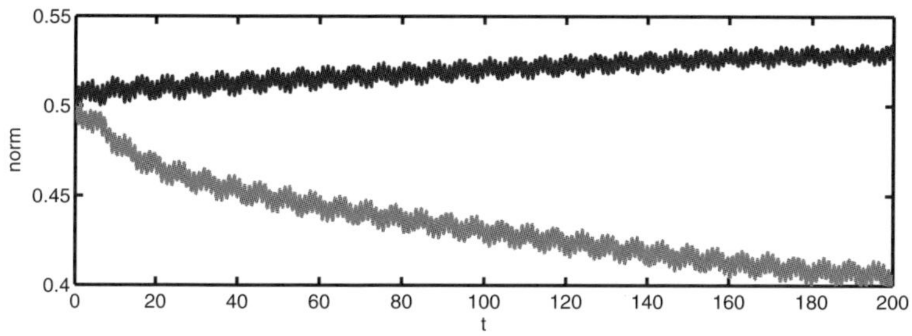

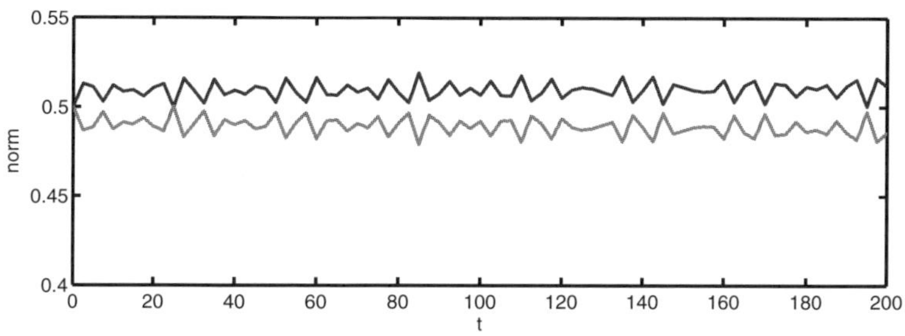

Figure 10. Trapped state's projections on ground and excited states. Top plot is for the case where $\omega_{res} = 2E_{1*} - E_{0*} > 0$, $\Gamma > 0$. Bottom plot is for the case where $\omega_{res} < 0$, $\Gamma = 0$

Remark 5.1. The detailed analysis indicates that if one considers initial data, which is a superposition of a nonlinear ground state and a nonlinear excited state, then half the excited state energy is radiated and half goes into to forming a new asymptotic ground state [40,41,45]: $|\alpha_0^{\pm}|^2 \sim |\alpha_0(0)|^2 + \frac{1}{2}|\alpha_1(0)|^2$; see (5.9).

Remark 5.2. This *ground state selection* has been observed in experiments in optical waveguides[31].

Remark 5.3. For related work on asymptotic behavior for NLS type equations see, for example, Refs. 5, 39, 6, 10, 11, 43, 44, 16, 45.

Remark 5.4. The "emission" of energy from the excited state into the ground state and dispersive radiation channels is a nonlinear variant of phenomena such as spontaneous emission, associated with the embedded eigenvalues in the continuous spectrum; see, for example, Refs. 33, 38, 12 and references cited therein. See also, related work on parametrically excited Hamiltonian systems – deterministic and random: Refs. 37, 28, 29, 30

Sketch of the analysis:
We view the full infinite dimensional Hamiltonian system (PDE) as being comprised of two weakly coupled subsystems:

- a finite dimensional (nonlinear oscillators - ODEs) governing the interacting nonlinear bound states (particles), and
- an infinite dimensional (wave equation - PDE) governing dispersive radiation.

We obtain this equivalent formulation beginning with the following Ansatz:

$$\Phi(x,t) \;=\; \Psi_{\alpha_0(t)} e^{-i\Theta_0(t)} \;+\; \Psi_{\alpha_1(t)} e^{-i\Theta_1(t)} \;+\; \eta_{\rm rad}(t) \tag{5.3}$$

The functions $\alpha_j(t)$ and $\Theta_j(t)$, $j = 0, 1$ are "collective coordinates" on nonlinear bound state manifolds of equilibria and $\eta_{\rm rad}(t)$ denotes dispersive radiation; see, for example, Refs. 46, 34, 35, 36, 4, 23.

Substitution into (5.1) and projecting with respect to an appropriate biorthogonal basis of the adjoint problem yields an equivalent system in terms of $\alpha_0(t), \alpha_1(t)$ and $\eta_{\rm res}(t)$ having the form of

Oscillators interacting with a field:

$$\begin{aligned} i\partial_t \alpha_0 &= \mathcal{C}_{\alpha_0}(\alpha_0, \alpha_1, \eta_{\rm rad}) \\ i\partial_t \alpha_1 &= \mathcal{C}_{\alpha_1}(\alpha_0, \alpha_1, \eta_{\rm rad}) \\ i\partial_t \eta_{\rm rad} &= H\eta_{\rm rad} \;+\; \mathcal{P}_c(H)\; \mathcal{R}[\alpha_0, \alpha_1, \eta_{\rm rad}] \end{aligned} \tag{5.4}$$

Approximate finite dimensional reduction: In a manner analogous to centre-manifold reduction of dissipative systems [8,24], we next attempt to find a closed system by approximately solving for the radiation components, $\eta_{\rm rad}$, as functional of oscillator variables, α_j. *In particular, we find the contributions responsible for resonant energy exchange between oscillator and field degrees of freedom.* These involve spectral components in a neighborhood of frequency $\omega_{res} \;=\; 2E_{1*} - E_{0*} > 0$:

$$\eta_{\rm rad} \;\sim\; \eta_{\rm rad}^{\rm res}[\alpha_0, \alpha_1].$$

We obtain a finite dimensional system, a set of ODEs in **normal form** [22], which captures, up to controllable corrections, the energy loss from the oscillators due to *radiation damping.* This normal form is weakly coupled to a dispersive wave equation, whose effect decreases with advancing time. For large time t, this effect can be estimated in the spirit of low energy scattering phenomena in the absence of coherent structures.

For concreteness, we illustrate the steps of the argument, beginning with a *model oscillator - field system*, closely related to our analysis:

$$\begin{aligned} i\partial_t A_0 &= \langle \chi, \overline{\eta(\cdot,t)} \rangle \; A_1^2 e^{-i\omega_{rest}t} \;+\; \ldots \\ i\partial_t A_1 &= 2\langle \chi, \eta(\cdot,t) \rangle \; \overline{A_1} A_0 e^{i\omega_{rest}t} + \ldots \\ i\partial_t \eta &= -\Delta\eta \;+\; \chi \overline{A_0} A_1^2 e^{-i\omega_{rest}t} + \ldots \\ \text{where } + \ldots &= \text{dispersive PDE corrections,} \end{aligned} \tag{5.5}$$

and χ denotes a spatially localized function.

The key contribution to the radiation field, due to resonance (because $0 < \omega_{res} \in \text{spec}(-\Delta) = [0, \infty)$) is

$$\eta_{\text{res}} \sim -i\pi\, e^{-i\omega_{res} t}\, \overline{A_0} A_1^2\, \Im\, (-\Delta - \omega_{res} - i0)^{-1} [\chi \cdot] + \ldots \tag{5.6}$$

Using (5.6) to approximately close the system for A_0 and A_1 yields the *dispersive normal form*

$$\begin{aligned} i\partial_t A_0 &= (\Lambda_0 + i\Gamma)\, |A_1|^4 A_0 + \ldots \\ i\partial_t A_1 &= (\Lambda_1 - 2i\Gamma)\, |A_0|^2 |A_1|^2 A_1 + \ldots, \quad \Gamma > 0 \end{aligned}$$

The precise character of the dynamics is made transparent if we introduce (renormalized) ground state and excited state energies:

$$P_0(t) \sim |A_0(t)|^2, \qquad P_1(t) \sim |A_1(t)|^2. \tag{5.7}$$

Here $\sim$ refers to equality up to near-identity change of variables. For sufficiently large times, $t_1(\Phi_0) \leq t$ we have the *nonlinear master equations:*

$$\frac{dP_0}{dt} \sim \Gamma P_1^2\, P_0 + \ldots, \qquad \frac{dP_1}{dt} \sim -2\Gamma P_1^2\, P_0 + \ldots, \quad \Gamma > 0 \tag{5.8}$$

From this we can show, for generic initial conditions, that as $t \to \pm\infty$ the system

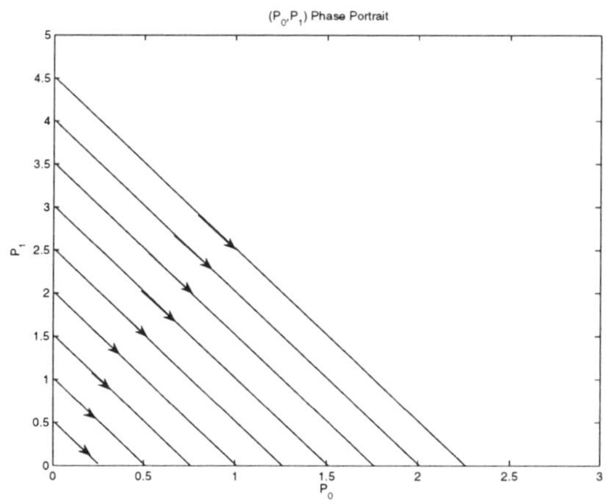

Figure 11. Phase portrait, corresponding to leading terms in (5.8), governing large t ground state selection

crystallizes on the ground state; see Figure 11. Furthermore, it follows from (5.8) that $2P_0(t) + P_1(t) \sim 2P_0(0) + P_1(0)$. Taking the limit as $t \to \infty$ and using the generic decay of $P_1(t)$ gives:

$$P_0(\infty) \;=\; P_0(0) \;+\; \frac{1}{2} P_1(0), \tag{5.9}$$

see Remark 5.1.

References

1. A.B. Aceves and S. Wabnitz. Self-induced transparency solitons in nonlinear refractive periodic media. *Phys. Lett. A*, 141:37–42, 1989.
2. R. Adami, C. Bardos, F. Golse, and A. Teta. Towards a rigorous derivation of the cubic nlse in dimension one. *Asymptot. Anal.*, 40:93–108, 2004.
3. R.W. Boyd. *Nonlinear Optics.* Academic Press, Boston, 2nd edition, 2003.
4. V.S. Buslaev and G.S. Perel'man. Scattering for the nonlinear Schrödinger equation: states close to a soliton. *St. Petersburg Math. J.*, 4:1111–1142, 1993.
5. V.S. Buslaev and G.S. Perel'man. On the stability of solitary waves for nonlinear Schrödinger equation. *Amer. Math. Soc. Transl. Ser. 2*, 164:75–98, 1995.
6. V.S. Buslaev and C. Sulem. On asymptotic stability of solitary waves for nonlinear Schrödinger equations. *Ann. Inst. H. Poincaré Anal. Non Linéaire*, 20:419–447, 2003.
7. X.D. Cao and B.A. Malomed. Soliton-defect collisions in the nonlinear Schrödinger equation. *Phys. Lett. A*, 206:177–182, 1995.
8. J. Carr. *Applications of Centre Manifold Theory.* Springer-Verlag, New York, 1981.
9. D.N. Christodoulides and R.I. Joseph. Slow Bragg solitons in nonlinear periodic structures. *Phys. Rev. Lett.*, 62:1746–1749, 1989.
10. S. Cuccagna. Stabilization of solutions to nonlinear Schrödinger equations. *Comm. Pure Appl. Math.*, 54(9):1110–1145, 2001.
11. S. Cuccagna. On asymptotic stability of ground states of nonlinear Schrödinger equations. *Rev. Math. Phys.*, 15, 2003.
12. S. Cuccagna. Spectra of positive and negative energies in the linearized NLS problem. *Commun. Pure Appl. Math.*, 58:1–29, 2005.
13. R. Dohnal and A.B. Aceves. Optical soliton bullets in (2+1)d nonlinear Bragg resonant periodic structures. *Stud. App. Math.*, 115:209–232, 2005.
14. B.J. Eggleton, C.M. de Sterke, and R.E. Slusher. Nonlinear pulse propagation in Bragg gratings. *J. Opt. Soc. Am B*, 14:2980–2992, 1997.
15. A. Elgart, L. Erdös, B. Schlein, and H-T Yau. The Gross-Pitaevskii equation as the mean field llimit of weakly coupled bosons. *Arch. Rat. Mech. Anal.*, 179:265–283, 2006.
16. Z. Gang and I.M. Sigal. On soliton dynamics in nonlinear Schrödinger equations. *arxiv:math-ph/0603059*, 2006.
17. R.H. Goodman, P.J. Holmes, and M.I. Weinstein. Nonlinear propagation of light in one-dimensional periodic structures. *J.Nonlinear Sci.*, 11:123–168, 2001.
18. R.H. Goodman, P.J. Holmes, and M.I. Weinstein. Strong NLS soliton-defect interactions. *Physica D*, 161(1):21–44, 2004.
19. R.H. Goodman, R.E. Slusher, and M.I. Weinstein. Stopping light on a defect. *J. Opt. Soc. Am. B*, 19:1635–1652, 2002.
20. R.H. Goodman, R.E. Slusher, and M.I. Weinstein. Trapping light pulses at controlled perturbations in periodic optical structures. *US Patent 6801685*, 2004.
21. R.H. Goodman and M.I. Weinstein. Stability of nonlinear defect states in the coupled mode equations, preprint. 2006.
22. J. Guckenheimer and P. Holmes. *Nonlinear Oscillations, Dynamical Systems and Bifurcations of Vector Fields.* Springer-Verlag, New York, 1983.
23. S. Gustafson, K. Nakanishi, and T-P. Tsai. Asymptotic stability and completeness in the energy space for nonlinear Schrödinger equations with small solitary waves. *IMRN*, (66):3559–3584, 2004.
24. D. Henry. *Geometric Theory of Semilinear Parabolic Equations.* Springer-Verlag, New York, 1981.
25. J. Holmer, J. Marzuola, and M. Zworski. Fast soliton scattering by delta impurities. *http://arxiv.org/pdf/math.AP/0602187*, 2006.

26. J. Holmer, J. Marzuola, and M. Zworski. Soliton splitting by external delta potentials. *preprint*, 2006.
27. J.L. Journé, A. Soffer, and C.D. Sogge. Decay estimates for Schrdinger operators. *Commun. Pure Appl. Math.*, 44:573–604, 1991.
28. E. Kirr and M.I. Weinstein. Parametrically excited Hamiltonian partial differential equations. *SIAM J. Math. Anal.*, 33:16–52, 2001.
29. E. Kirr and M.I. Weinstein. Metastable states in parametrically excited multimode hamiltonian partial differential equations. *Commun. Math. Phys.*, 236:335–372, 2003.
30. E. Kirr and M.I. Weinstein. Diffusion of power in randomly perturbed hamiltonian partial differential equations. *Commun. Math. Phys.*, 255:293–328, 2005.
31. D. Mandelik, Y. Lahini, and Y. Silberberg. Nonlinear induced relaxation to the ground state in a two-level system. *Phys. Rev. Lett.*, 95:073902, 2005.
32. H.A. Rose and M.I. Weinstein. On the bound states of the nonlinear Schrödinger equation with a linear potential. *Physica D*, 30:207–218, 1988.
33. I.M. Sigal. Nonlinear wave and schrödinger equations i. instability of time-periodic and quasiperiodic solutions. *Commun. Math. Phys.*, 153:297, 1993.
34. A. Soffer and M.I. Weinstein. Multichannel nonlinear scattering in nonintegrable systems. In *Lecture Notes in Physics: Integrable Systems and Applications*, volume 342, Berlin, 1989. Springer-Verlag.
35. A. Soffer and M.I. Weinstein. Multichannel nonlinear scattering in nonintegrable systems. *Commun. Math. Phys.*, 133:119–146, 1990.
36. A. Soffer and M.I. Weinstein. Multichannel nonlinear scattering and stability ii. the case of anisotropic potentials and data. *J. Diff. Eqns*, 98:376–390, 1992.
37. A. Soffer and M.I. Weinstein. Nonautonomous Hamiltonians. *J. Stat. Phys.*, 93:359–391, 1998.
38. A. Soffer and M.I. Weinstein. Time dependent resonance theory. *Geom. Func. Anal.*, 8:1086–1128, 1998.
39. A. Soffer and M.I. Weinstein. Resonances, radiation damping and instability of Hamiltonian nonlinear waves. *Invent. Math.*, 136:9–74, 1999.
40. A. Soffer and M.I. Weinstein. Selection of the ground state in nonlinear Schrödinger equations. *Rev. Math. Phys.*, 16(16):977–1071, 2004.
41. A. Soffer and M.I. Weinstein. Theory of nonlinear dispersive waves and selection of the ground state. *Phys. Rev. Lett.*, 95:213905, 2005.
42. C. Sulem and P.L. Sulem. *The Nonlinear Schrödinger Equation.* Springer, New York, 1999.
43. T.-P. Tsai and H.-T. Yau. Asymptotic dynamics of nonlinear Schrödinger equations: resonance dominated and dispersion dominated solutions. *Commun. Pure Appl. Math.*, 55:0153–0216, 2002.
44. T.-P. Tsai and H.-T. Yau. Relaxation of excited states in nonlinear Schrödinger equations. *Int. Math. Res. Not.*, 31:1629–1673, 2002.
45. M. I. Weinstein. Extended Hamiltonian Systems. In *Handbook of Dynamical Systems*, pages 1135–153, Amsterdam, 2006. Elsevier B.B.
46. M.I. Weinstein. Modulational stability of ground states of nonlinear Schrödinger equations. *SIAM J. Math. Anal.*, 16:472–491, 1985.
47. K. Yajima. The $W^{k,p}$ continuity of wave operators for Schrödinger operators. *J.. Math. Soc. Japan*, 47:551–581, 1995.

SOME MATHEMATICAL PROPERTIES OF LONG WAVES

D. J. BENNEY

Department of Mathematics, Massachusetts Institute of Technology, Cambridge, MA 02139, USA
E-mail: djb@math.mit.edu

Some aspects of long surface waves are considered. Special attention is focused on the existence of conservation laws for this physical system and related problems.

Keywords: Long wave; Conservation law; Moment; Nonlinear.

1. Introduction

The classical problem of wave propagation at a free surface has a long history and has provided motivation for the development of many methods used in applied mathematics.

For such problems there are three well know theoretical regimes. These are most simply identified in terms of three length scales, a_0 the wave amplitude, h_0 the mean depth and l_0 the wave length. The two dimensionless parameters $\epsilon = a_0/h_0$ and $\mu = h_0/l_0$ these theories correspond to the following limits

(i) Quasilinear theory, $\epsilon \ll 1, \mu \sim 1$. Here resonances and energy exchanges are of primary interest.
(ii) Nonlinear long waves, $\epsilon \sim 1, \mu \ll 1$. The theory predicts wave breaking and leads to studies of hyperbolic systems.
(iii) Intermediate theory, $\epsilon \sim \mu^2 \ll 1$. Here weak dispersion and nonlinearity are in balance and lead to the KdV equation, where an infinity of conservation laws and soliton properties exist.

2. Two-dimensional analysis

Consider the classical problem of time-dependent motion on an inviscid fluid of constant density ρ under the action of gravity g. In two dimensions let $y = 0$ be the rigid bottom and $y = h(x,t)$ the free surface. The two velocity components satisfy the Euler equations, namely,

$$\frac{\partial u}{\partial x}+\frac{\partial v}{\partial y}=0, \tag{1}$$

$$\frac{\partial u}{\partial t}+u\frac{\partial u}{\partial x}+v\frac{\partial u}{\partial y}=-\frac{1}{\rho}\frac{\partial p}{\partial x}, \tag{2}$$

$$\frac{\partial v}{\partial t}+u\frac{\partial v}{\partial x}+v\frac{\partial v}{\partial y}=-\frac{1}{\rho}\frac{\partial p}{\partial y}-g. \tag{3}$$

The appropriate boundary conditions are that

$$v=0, \qquad y=0, \tag{4}$$

$$p=p_0, \qquad y=h, \tag{5}$$

$$\frac{\partial h}{\partial t}+u\frac{\partial h}{\partial x}-v=0, \qquad y=h, \tag{6}$$

where p_0 is the constant atmospheric pressure.

In this study interest is focused on nonlinear long wave theory (sometimes referred to as nonlinear shallow water theory), and it is assumed that the ratio of vertical to horizontal scales is a small parameter, but the wave amplitude is arbitrary. It is well known that the consequence of these assumptions being applied to the preceding problem is the following simplified system:

$$\frac{\partial u}{\partial x}+\frac{\partial v}{\partial y}=0, \tag{7}$$

$$\frac{\partial u}{\partial t}+u\frac{\partial u}{\partial x}+v\frac{\partial u}{\partial y}=-g\frac{\partial h}{\partial x} \tag{8}$$

subject to the boundary conditions

$$v=0, \qquad y=0, \tag{9}$$

$$\frac{\partial h}{\partial t}+u\frac{\partial h}{\partial x}-v=0, \qquad y=h(x,t). \tag{10}$$

This new initial value problem remains difficult, save in the very special (classical) case where

$$u=u(x,t), \qquad v=-y\frac{\partial u}{\partial x} \tag{11}$$

in which case the following hyperbolic system for u and h is obtained:

$$\frac{\partial u}{\partial t}+u\frac{\partial u}{\partial x}+g\frac{\partial h}{\partial x}=0, \tag{12}$$

$$\frac{\partial h}{\partial t}+u\frac{\partial h}{\partial x}+h\frac{\partial u}{\partial x}=0. \tag{13}$$

These equations predict wave breaking so that the theory eventually fails, unless some dispersion or dissipation is included.

For the more general case where $u(x,y,0)$ and $h(x,0)$ are prescribed, the flow may be locally unstable and possible motion becomes much more complex. However, the general two-dimensional system does have some interesting properties. These are seen if Eq. (8) is multiplied by u^{n-1} and integrated with respect to y from $y = 0$ to $y = h$. The resulting equation is

$$\frac{\partial A_n}{\partial t} + \frac{\partial A_{n+1}}{\partial x} + gnA_{n-1}\frac{\partial A_0}{\partial x} = 0, \qquad n \geq 1, \tag{14}$$

where moments A_r are defined by

$$A_r(x,t) = \int_0^h u^r(x,y,t)\mathrm{d}y, \qquad r \geq 0. \tag{15}$$

Laws corresponding to conservation of mass, momentum and energy are readily found to be

$$\frac{\partial A_0}{\partial t} + \frac{\partial A_1}{\partial x} = 0, \tag{16}$$

$$\frac{\partial A_1}{\partial t} + \frac{\partial}{\partial x}\left(A_2 + \frac{1}{2}gA_0^2\right) = 0, \tag{17}$$

$$\frac{\partial}{\partial t}\left(A_2 + gA_0^2\right) + \frac{\partial}{\partial x}\left(A_3 + 2gA_0A_1\right) = 0 \tag{18}$$

and as has been shown by Benney[1] there are an infinity of such moment conservation laws. Additionally Miura[2] has proved that there exist an infinity of local conservation laws.

In dealing with the moment conservation laws it is convenient to use either of the generalizing functions F or G where

$$F(x,t;\alpha) = \sum_{n=0}^{\infty} A_n\alpha^n = \int_0^h \frac{\mathrm{d}y}{1-\alpha u}, \tag{19}$$

$$G(x,t;\alpha) = \sum_{n=0}^{\infty} \frac{A_n\alpha^n}{n!} = \int_0^h e^{\alpha u}\mathrm{d}y. \tag{20}$$

The relevant equations for F and G are found to be

$$\alpha\frac{\partial F}{\partial t} + \frac{\partial F}{\partial x} = \left\{1 - g\alpha^2\frac{\partial}{\partial \alpha}(\alpha F)\right\}\frac{\partial F}{\partial x}(x,t;0), \tag{21}$$

$$\frac{\partial G}{\partial t} + \frac{\partial^2 G}{\partial \alpha \partial x} + g\alpha G\frac{\partial G}{\partial x}(x,t;0) = 0. \tag{22}$$

3. An analogous mathematical system

A slight mathematical extension for the sequential moments considered earlier is the infinite system

$$\frac{\partial C_n}{\partial t} + a_n\frac{\partial C_{n+1}}{\partial x} + gb_nC_{n-1}\frac{\partial C_0}{\partial x} = 0, \qquad n \geq 0, \tag{23}$$

where a_n and b_n are prescribed functions of n. The question of concern is whether this more general system has an infinity of conservation laws. With this purpose in mind, it is instructive to attempt to derive some of the low order laws. Not unexpectedly the first three laws are

$$\frac{\partial C_0}{\partial t} + a_0 \frac{\partial C_1}{\partial x} = 0, \tag{24}$$

$$\frac{\partial C_1}{\partial t} + \frac{\partial}{\partial x}\left(a_1 C_2 + \frac{g b_1}{2} C_0^2 \right) = 0, \tag{25}$$

$$\frac{\partial}{\partial t}\left(C_2 + \frac{g b_2}{2a_0} C_0^2 \right) + \frac{\partial}{\partial x}(a_2 C_3 + g b_2 C_0 C_1) = 0. \tag{26}$$

Less obvious is the fact that two further conservation laws exist. These are

$$\frac{\partial}{\partial t}\left(C_3 + \frac{g b_3}{a_1} C_0 C_2 \right) + \frac{\partial}{\partial x}\left(a_3 C_4 + g b_3 C_0 C_2 + \frac{g a_0 b_3}{2a_1} C_1^2 + \frac{g^2 b_1 b_3}{3a_1} C_0^3 \right) = 0, \tag{27}$$

$$\begin{aligned} &\frac{\partial}{\partial t}\left(C_4 + \frac{g b_4}{a_2} C_0 C_2 + \frac{g b_4 a_0}{2 a_1 a_2} C_1^2 + \left\{ \frac{g^2 b_2 b_4}{6 a_0 a_2} + \frac{g^2 b_4}{2 a_1 a_2} \right\} C_0^3 \right) \\ &\quad + \frac{\partial}{\partial x}\left(a_4 C_5 + g b_4 C_0 C_3 + \frac{g b_4 a_0}{a_2} C_1 C_2 + \left\{ \frac{g^2 b_2 b_4}{2a_2} + \frac{g^2 a_0 b_4}{2 a_1 a_2} \right\} C_1 C_0^2 \right) = 0. \end{aligned} \tag{28}$$

However, without constraints on the constants a_n and b_n, no further laws of moment type appear to be possible. For example, an attempt for a sixth law yields

$$\begin{aligned} &\frac{\partial}{\partial t}\left(C_5 + \frac{g b_5}{a_3} C_0 C_3 + \frac{g a_0 b_5}{a_2 a_3} C_1 C_2 - \frac{g^2 b_2 b_5}{a_1 a_3} C_0^2 C_1 \right) \\ &\quad + \frac{\partial}{\partial x}\left(a_5 C_6 + g b_5 C_0 C_4 + \frac{g a_0 b_5}{a_3} C_1 C_3 + \frac{g a_0 a_1 b_5}{2 a_2 a_3} C_2^2 \right. \\ &\quad \left. + \frac{g^2 a_0 b_2}{a_2 a_3} C_0 C_1^2 + \frac{g^2 b_5}{2a_3}\left(b_3 + \frac{a_0 b_1}{a_2} \right) C_0^2 C_2 + \frac{g^3 a_0 b_1^2 b_5}{8 a_1 a_2 a_3} C_0^4 \right) + K_6 C_0^2 \frac{\partial C_1}{\partial t} = 0, \end{aligned} \tag{29}$$

where

$$K_6 = \frac{g^2 b_5}{a_3}\left[\frac{a_0 b_1}{2 a_1 a_2} - \frac{b_2}{a_2} + \frac{b_3}{2a_1} \right]. \tag{30}$$

Unless $K_6 = 0$ the system has only five conservation laws. Even if $K_6 = 0$, additional constraints are to be anticipated at higher moment levels in order to preserve conservation properties. Note that for the A_n defined by Eq. (19), $a_n = 1$ and $b_n = n$ while for those defined by (20), $a_n = n + 1$ and $b_n = 1$. In each case $K_6 = 0$. Indeed, it is a simple matter to show that the condition

$$\frac{a_n b_{n+1}}{n+1} = \text{constant} \tag{31}$$

is sufficient to ensure an infinity of conservation laws.

4. Three-dimensional analysis

Analogous mathematical issues arise in higher dimensions. More complicated physical mechanisms might be expected to limit conservation properties. Indeed, this seems to be the case.

In three dimensions, with z as the second horizontal coordinate, the nonlinear long wave equations are

$$\frac{\partial u}{\partial x} + \frac{\partial v}{\partial y} + \frac{\partial w}{\partial z} = 0, \tag{32}$$

$$\frac{\partial u}{\partial t} + u\frac{\partial u}{\partial x} + v\frac{\partial u}{\partial y} + w\frac{\partial u}{\partial z} = -g\frac{\partial h}{\partial x}, \tag{33}$$

$$\frac{\partial w}{\partial t} + u\frac{\partial w}{\partial x} + v\frac{\partial w}{\partial y} + w\frac{\partial w}{\partial z} = -g\frac{\partial h}{\partial z} \tag{34}$$

with boundary conditions

$$v = 0, \qquad y = 0 \tag{35}$$

$$\frac{\partial h}{\partial t} + u\frac{\partial h}{\partial x} + w\frac{\partial h}{\partial z} - v = 0, \qquad y = h(x, z, t). \tag{36}$$

Using the moments $A_{r,s}$ defined by

$$A_{r,s} = \int_0^h u^r w^s \mathrm{d}y \tag{37}$$

leads to the evolution equations

$$\frac{\partial A_{r,s}}{\partial t} + \frac{\partial A_{r+1,s}}{\partial x} + \frac{\partial A_{r,s+1}}{\partial z} + grA_{r-1,s}\frac{\partial A_{0,0}}{\partial x} + gsA_{r,s-1}\frac{\partial A_{0,0}}{\partial z} = 0. \tag{38}$$

Once again mass, momentum (z) and energy conservation give the four laws

$$\frac{\partial A_{0,0}}{\partial t} + \frac{\partial A_{1,0}}{\partial x} + \frac{\partial A_{0,1}}{\partial z} = 0, \tag{39}$$

$$\frac{\partial A_{1,0}}{\partial t} + \frac{\partial}{\partial x}\left(A_{2,0} + \frac{gA_{0,0}^2}{2}\right) + \frac{\partial A_{1,1}}{\partial z} = 0, \tag{40}$$

$$\frac{\partial A_{0,1}}{\partial t} + \frac{\partial A_{1,1}}{\partial x} + \frac{\partial}{\partial z}\left(A_{0,2} + \frac{gA_{0,0}^2}{2}\right) = 0, \tag{41}$$

$$\frac{\partial}{\partial t}\left(A_{2,0} + A_{0,2} + gA_{0,0}^2\right) + \frac{\partial}{\partial x}\left(A_{3,0} + A_{1,2} + 2gA_{1,0}A_{0,0}\right) + \frac{\partial}{\partial z}\left(A_{0,3} + A_{2,1} + 2gA_{0,1}A_{0,0}\right) = 0. \tag{42}$$

However, attempts by the author to find other moment conservation laws lead him to believe that there are no more.

References

1. D. J. Benney, Stud. Appl. Math. **52**, 45 (1973).
2. R. M. Miura, Stud. Appl. Math. **53**, 45 (1974).

NEW SOLITARY WAVE STRUCTURES IN TWO-DIMENSIONAL PERIODIC MEDIA

ZUOQIANG SHI* AND JIANKE YANG†
Zhou Pei-Yuan Center for Applied Mathematics
Tsinghua University
Beijing, China, 100084

New solitary-wave structures in two-dimensional periodic media are obtained in the context of a two-dimensional nonlinear Schrödinger equation with a periodic potential. These new structures bifurcate from the edges of Bloch bands with two linearly independent Bloch modes. Away from these band edges, superposition of these Bloch modes, modulated by nonlinear effects, give rise to composite solitary waves with distinctive intensity and phase profiles such as vortex arrays. Using perturbation methods, coupled nonlinear envelope equations for the two Bloch waves near the band edges are analytically derived. Numerically, these composite solitons are directly computed both near and far away from the band edges, and the analytical results are fully confirmed.

1. Introduction

Nonlinear wave propagation in periodic media is attracting a lot of attention these days. This was stimulated in part by rapid advances in optics, Bose-Einstein condensation, and related fields. In optics, various periodic and quasi-periodic structures (such as photonic crystals, photonic crystal fibers, periodic waveguide arrays and photonic lattices) have been constructed by ingeneous experimental techniques, with applications to light routing, switching and optical information processing [1,2,3,4,5,6]. This periodic medium creates a wide range of new phenomena for light propagation, even in the linear regime. For instance, the diffraction of light in a periodic medium exhibits distinctively different patterns from homogeneous diffraction [3]. If the periodic medium has a local defect, this defect can guide light by a totally new physical mechanism called repeated Bragg reflections [2,7,8]. When the nonlinear effects become significant, say with high-power beams or in strongly nonlinear materials such as photorefractive crystals, the physical phenomena would be even richer and more complex, and their understanding is far from complete yet. In Bose-Einstein condensates, one direction of recent research is to load these condensates into periodic optical lattices [9]. This problem and the above nonlinear optics problems are closely

*email: shizq03@mails.tsinghua.edu.cn
†Permanent address: Department of Mathematics and Statistics, University of Vermont, Burlington, VT 05401, USA; email: jyang@math.uvm.edu.

related, and are often analyzed together in the mathematical community.

Solitary waves play an important role in nonlinear wave systems. These waves are nonlinear localized structures which propagate without change of shape. In physical communities, they are often just called solitons, which we do occasionally in this paper as well. If a physical system admits solitary waves, it often has important consequences. For instance, optical fibers can support solitary waves (pulses) when the nonlinearity of the pulse balances linear dispersion. This fact led to soliton-based fiber communication systems, which greatly propelled the telecommunication industry in the end of the last century. In one-dimensional periodic media, solitary waves (called lattice solitons) also exist, and they have been observed in optical experiments [3,10,11,12]. But two- (or higher-) dimensional periodic media can support a much wider array of solitary wave structures which have no counterpart in one-dimensional systems. One example is the vortex lattice solitons which was predicted in [13,14] and later observed in [15,16]. These vortex solitons lie in the semi-infinite bandgap of the periodic system. Recently, Bartal, et al. [17] reported the observation of vortex solitons which lie in a higher bandgap of a periodic medium, and Makazyuk, et al. [18] reported the observation of linear localized light patterns which comprise of dipole or vortex-cell arrays in a defected 2D lattice. Even though these two observations are quite different, their common feature is that Bloch-wave superpositions were essential for their explanations. These observations indicate that Bloch-wave superpositions can create novel and intricate solitary-wave patterns. However, it is not clear at the moment what are all the possible solitary-wave structures this Bloch-wave superposition can create, and which edges of Bloch bands these structures can bifurcate from. In-depth analytical studies on these new structures are totally missing as well.

In this paper, we analyze all possible solitary-wave structures due to Bloch-wave superpositions in two-dimensional periodic media both analytically and numerically, using the two-dimensional nonlinear Schrödinger equation with a periodic potential as the mathematical model. First, we identify edges of Bloch bands which admit two linearly independent Bloch modes. Then using perturbation methods, we derive coupled nonlinear equations for envelopes of these superposing Bloch waves near these band edges. We find that these envelope equations admit solutions which lead to novel solitary-wave structures such as vortex-array solitons if the nonlinearity is of the same sign as the second-order dispersion coefficients of Bloch waves on the underlying band edges. Hence these composite solitons exist for both focusing and defocusing nonlinearities. We have also computed these composite solitons directly by numerical methods both near and further away from the band edges. In the former case, the numerical results are in full agreement with the analytical calculations.

2. The Mathematical Model

The mathematical model for the study of solitary waves in periodic media is the two-dimensional (2D) nonlinear Schr"odinger (NLS) equation with a periodic potential:

$$iU_t + U_{xx} + U_{yy} - V(x,y)U + \sigma|U|^2U = 0, \tag{1}$$

where $V(x,y)$ is the periodic potential, and $\sigma = \pm 1$ is the sign of nonlinearity. This model naturally arises for light propagation in a periodic Kerr medium, and for Bose-Einstein condensates trapped in an optical lattice [9]. In certain optical materials (such as photorefractive crystals), the nonlinearity is of a different (saturable) type. But it is known that these different types of nonlinearities give qualitatively similar results [14,15,16,19].

In this article we take the periodic potential as

$$V(x,y) = V_0\left(\sin^2 x + \sin^2 y\right), \tag{2}$$

whose periods along the x and y directions are both equal to π. This potential is separable, which facilitates our theoretical analysis. In addition, without loss of generality, when specific computations are carried out, we always set $V_0 = 6$.

Solitary waves in Eq. (1) are sought in the form

$$U(x,y,t) = u(x,y)e^{-i\mu t}, \tag{3}$$

where amplitude function $u(x,y)$ is a solution of the following equation:

$$u_{xx} + u_{yy} - [F(x) + F(y)]u + \mu u + \sigma|u|^2u = 0, \tag{4}$$

$$F(x) = V_0 \sin^2 x, \tag{5}$$

and μ is a propagation constant.

3. Bloch bands and band gaps

When function $u(x,y)$ is infinitesimal, Eq. (4) becomes a linear equation:

$$u_{xx} + u_{yy} - [F(x) + F(y)]u + \mu u = 0. \tag{6}$$

Solutions of this linear equation are the Bloch modes, and the corresponding propagation constants μ form Bloch bands. Since the potential in (6) is separable, Bloch solutions and Bloch bands of the 2D equation (6) can be obtained from solutions of a 1D equation. Specifically, the 2D Bloch solution $u(x,y)$ of Eq. (6) and the propagation constant μ can be split into the following form:

$$u(x,y) = p(x;\omega_a)p(y;\omega_b), \quad \mu = \omega_a + \omega_b, \tag{7}$$

where $p(x;\omega)$ is a solution of the following 1D equation:

$$p_{xx} - F(x)p + \omega p = 0. \tag{8}$$

Eq. (8) is equivalent to the Mathieu equation. Its solution is

$$p(x;\omega) = e^{ikx}\tilde{p}(x;\omega), \tag{9}$$

where $\tilde{p}(x;\omega)$ is periodic with period π, and

$$\omega = \omega(k) \tag{10}$$

is the 1D dispersion relation. This dispersion diagram is shown in Fig. 1(a) (for $V_0 = 6$). The bandgap structure at various values of V_0 are shown in Fig. 1(b). The four Bloch waves at both edges of the lowest two Bloch bands are displayed in Fig. 2.

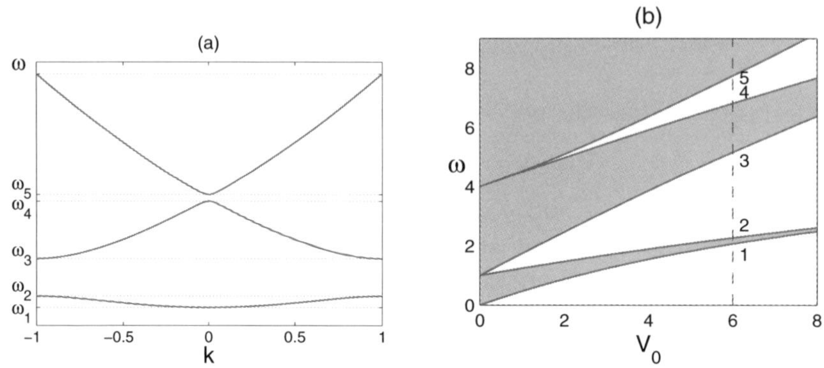

Figure 1. (a) Dispersion curves of the 1D equation (8) with $V_0 = 6$; (b) Bloch bands (shaded regions) and bandgaps at various values of potential levels V_0 in the 1D equation (8).

Using these 1D dispersion results and the above connection between 1D and 2D Bloch solutions, we can construct the dispersion surfaces and band gap structures for the 2D problem (6), and the results are shown in Fig. 3. In Fig. 3 (a), the dispersion surfaces of the 2D problem are displayed at $V_0 = 6$. The 2D bandgap structure at various values of V_0 are shown in Fig. 3 (b). Unlike the 1D case, for a given V_0 value, there are only a finite number of bandgaps in the 2D problem. The first bandgap appears only when $V_0 > 1.40$, the second bandgap appearing when $V_0 > 4.13$, etc. As V_0 increases further, more bandgaps will be found.

Now we examine the 2D Bloch solutions on the edges of Bloch bands. To illustrate, we consider the points A, B, C, D and E marked in Fig. 3 (b), where $V_0 = 6$. At each of points A and B, there is a *single* Bloch solution, which is symmetric along the x and y directions, i.e., $u(x,y) = u(y,x)$. In view of the relation (7), it is easy to see that the Bloch solution at point A is $p(x;\omega_1)p(y;\omega_1)$, where ω_1 marked in Fig. 1(a), and $p(x;\omega_2)$ is shown in Fig. 2(a). For convenience, we denote point A as "1 + 1". Similarly, the 2D Bloch solution at point B is $p(x;\omega_2)p(y;\omega_2)$, where ω_2 marked in Fig. 1(a), and $p(x;\omega_2)$ is shown in Fig. 2(b). Point B is $2+2$ in our notations. Points C, D, E are different from A and B and are much more interesting, however. At these points, there are *two* linearly independent Bloch solutions, $u(x,y)$ and $u(y,x)$. For instance, at point C, these two solutions are $p(x;\omega_1)p(y;\omega_3)$ and $p(y;\omega_1)p(x;\omega_3)$, where ω_3 is marked in Fig. 1(a), and $p(x;\omega_3)$

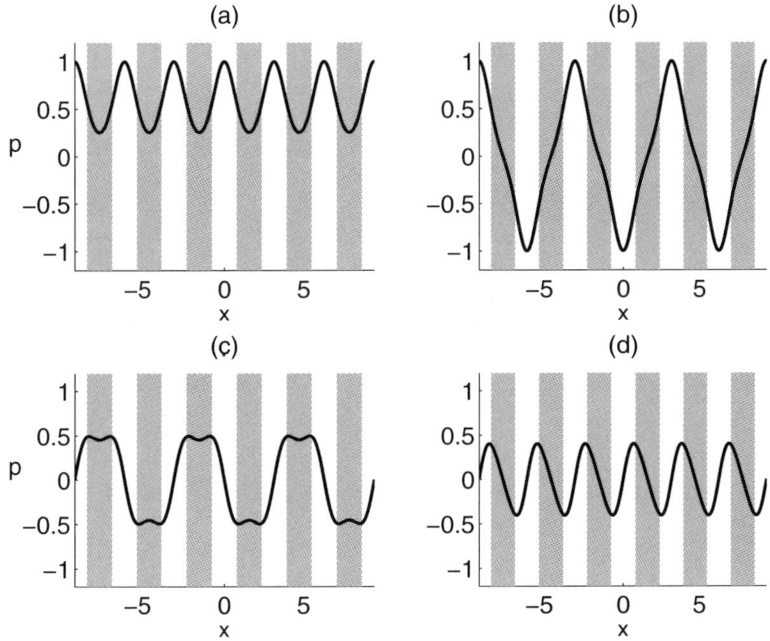

Figure 2. The first four 1D Bloch waves at edges of Bloch bands marked by letters $\omega_1, \omega_2, \omega_3$ and ω_4 in Fig. 1(a). (a): $\omega_1 = 2.063182$; (b): $\omega_2 = 2.266735$; (c): $\omega_3 = 5.165940$; (d): $\omega_4 = 6.81429$.

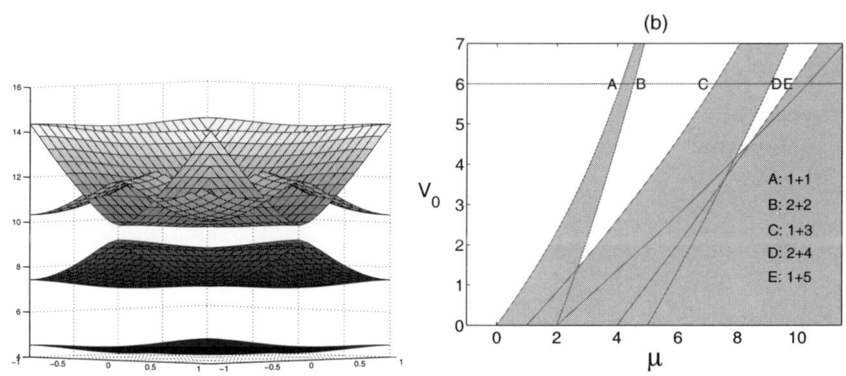

Figure 3. (a) Dispersion surfaces of the 2D problem (6) at $V_0 = 6$; (b) the 2D bandgap structure for various values of V_0.

is shown in Fig. 2(c). Point C is "1+3". The solution $p(x;\omega_1)p(y;\omega_3)$ is displayed in Fig. 4(a). Point D is "2+4", where the two linearly independent Bloch solutions are $p(x;\omega_2)p(y;\omega_4)$ and $p(y;\omega_2)p(x;\omega_4)$, the former of which is displayed in Fig. 4(b). Point E is "1+5". Because of the existence of two linearly independent Bloch solutions, their linear superposition remains a solution. These superpositions can give rise to interesting composite patterns such as vortex arrays, as has been pointed out in [17,18].

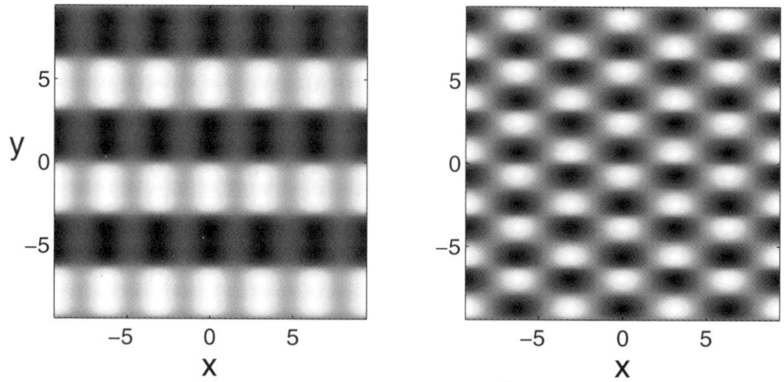

Figure 4. (a) A 2D Bloch mode at point C marked in Fig. 3; (b) A 2D Bloch mode at point D marked in Fig. 3.

The above superposing Bloch solutions exist on band edges with infinitesimal amplitudes. When amplitudes of these solutions increase, these Bloch solutions may localize and form solitary-wave structures. These solitary waves exist not on band edges, but inside bandgaps. In the next section, we analyze these solutions by perturbation methods in the limit when these solutions lie near the Bloch bands.

4. Asymptotic derivation of coupled envelope equations

In this section, we develop an asymptotic theory to analyze small-amplitude solitary waves bifurcating from superimposing Bloch waves near band edges in Eq. (4).

Let us consider a 2D band edge $\mu_0 = \omega_{0,1} + \omega_{0,2}$, where $\omega_{0,n}$ $(n = 1, 2)$ are 1D band edges, two linearly independent Bloch modes $p_1(x)p_2(y)$ and $p_1(y)p_2(x)$ exist at the edge, and $p_n(x) = p(x; \omega_{0,n})$ $(n = 1, 2)$. Notice that

$$p_n(x + L) = \pm p_n(x) \tag{11}$$

since $\omega_{0,n}$ is a 1D band edge, and $L = \pi$ is the period of the potential $F(x)$. We take an infinitesimal solution $u(x, y)$ of Eq. (4) which is a linear superposition of these two Bloch modes. When $u(x, y)$ is small but not infinitesimal, we can expand the solution $u(x, y)$ of Eq. (4) into a multi-scale perturbation series:

$$u = \epsilon u_0 + \epsilon^2 u_1 + \epsilon^3 u_2 + \cdots , \tag{12}$$

$$\mu = \mu_0 + \eta\epsilon^2, \tag{13}$$

where

$$u_0 = A_1(X, Y)p_1(x)p_2(y) + A_2(X, Y)p_2(x)p_1(y), \tag{14}$$

$\eta = \pm 1$, and $X = \epsilon x, Y = \epsilon y$. Substituting the above expansions into Eq. (4), the equation at $O(\epsilon)$ is automatically satisfied. At order $O(\epsilon^2)$, the equation is

$$u_{1xx} + u_{1yy} - (F(x) + F(y))u_1 + \mu_0 u_1 = -2\left(\frac{\partial^2 u_0}{\partial x \partial X} + \frac{\partial^2 u_0}{\partial y \partial Y}\right). \tag{15}$$

Its homogeneous equation has two linearly independent solutions, $p_1(x)p_2(y)$, and $p_1(y)p_2(x)$. In order for the inhomogeneous equation (15) to admit a solution, the following Fredholm conditions

$$\int_0^{2L}\int_0^{2L}\left(\frac{\partial^2 u_0}{\partial x \partial X} + \frac{\partial^2 u_0}{\partial y \partial Y}\right) p_1(x)p_2(y)\, dxdy = 0, \tag{16}$$

$$\int_0^{2L}\int_0^{2L}\left(\frac{\partial^2 u_0}{\partial x \partial X} + \frac{\partial^2 u_0}{\partial y \partial Y}\right) p_2(x)p_1(y)\, dxdy = 0, \tag{17}$$

must be satisfied. Here the integration length is $2L$ rather than L since the homogeneous solutions $p_1(x)p_2(y)$ and $p_2(x)p_1(y)$ may have periods $2L$ along the x and y direction (see Eq. (11)). It is easy to check that these conditions are indeed satisfied automatically, thus we can find a solution for Eq. (15) as

$$u_1 = \frac{\partial A_1}{\partial X}\nu_1(x)p_2(y) + \frac{\partial A_1}{\partial X}\nu_2(y)p_1(x) + \frac{\partial A_2}{\partial X}\nu_2(x)p_1(y) + \frac{\partial A_2}{\partial X}\nu_1(y)p_2(x), \tag{18}$$

where $\nu_n(x)$ is a periodic solution of equation

$$\nu_{n,xx} - F(x)\nu_n + \omega_{0,n}\nu_n = -2p_{n,x}, \quad n = 1, 2. \tag{19}$$

At $O(\epsilon^3)$, the equation is

$$u_{2xx} + u_{2yy} - (F(x) + F(y))u_2 + \mu_0 u_2 = \\ -\left(2\frac{\partial^2 u_1}{\partial x \partial X} + 2\frac{\partial^2 u_1}{\partial y \partial Y} + \frac{\partial^2 u_0}{\partial X^2} + \frac{\partial^2 u_0}{\partial Y^2} + \eta u_0 + |u_0|^2 u_0\right). \tag{20}$$

Substituting the expressions for u_0 and u_1 into this equation, we get

$$- [u_{2xx} + u_{2yy} - (F(x) + F(y))u_2 + \mu_0 u_2] =$$

$$\frac{\partial^2 A_1}{\partial X^2}\left(2\frac{\nu_1(x)}{\partial x} + p_1(x)\right)p_2(y) + \frac{\partial^2 A_1}{\partial Y^2}\left(2\frac{\nu_2(y)}{\partial y} + p_2(y)\right)p_1(x) + \\ \frac{\partial^2 A_2}{\partial X^2}\left(2\frac{\nu_2(x)}{\partial x} + p_2(x)\right)p_1(y) + \frac{\partial^2 A_2}{\partial Y^2}\left(2\frac{\nu_1(y)}{\partial y} + p_1(y)\right)p_2(x) + \\ 2\frac{\partial^2 A_1}{\partial X \partial Y}\left(\frac{\partial p_1(x)}{\partial x}\nu_2(y) + \frac{\partial p_2(y)}{\partial y}\nu_1(x)\right) + 2\frac{\partial^2 A_2}{\partial X \partial Y}\left(\frac{\partial p_2(x)}{\partial x}\nu_1(y) + \frac{\partial p_1(y)}{\partial y}\nu_2(x)\right) + \\ p_1^3(x)p_2^3(y)|A_1|^2 A_1 + p_1(x)p_2(y)p_1^2(y)p_2^2(x)(\bar{A}_1 A_2^2 + 2A_1|A_2|^2) + \\ p_2^3(x)p_1^3(y)|A_2|^2 A_2 + p_1^2(x)p_2^2(y)p_1(y)p_2(x)(\bar{A}_2 A_1^2 + 2A_2|A_1|^2) + \\ \eta A_1(X, Y, T)p_1(x)p_2(y) + \eta A_2(X, Y, T)p_2(x)p_1(y). \tag{21}$$

Here the overbar represents complex conjugation. Before applying the Fredholm conditions to this inhomogeneous equation, we notice the following identities:

$$\int_0^{2L} p_1(x)p_2(x)dx = 0, \tag{22}$$

$$\int_0^{2L} p_1(x)\nu_2(x)dx = \int_0^{2L} p_2(x)\nu_1(x)dx, \tag{23}$$

and

$$\int_0^{2L} \left(2\frac{\nu_n(x)}{\partial x} + p_n(x)\right) p_n(x)dx = D_n \int_0^{2L} p_n^2(x)dx, \quad n = 1, 2, \tag{24}$$

where

$$D_n \equiv \frac{1}{2}\frac{d^2\omega}{dk^2}\bigg|_{\omega=\omega_{0,n}}. \tag{25}$$

Identity (22) holds since $p_1(x)$ and $p_2(x)$ are the eigenfunctions of the self-adjoint linear Schr"odinger operator with different eigenvalues. Identity (23) can be confirmed by taking the inner product between Eq. (19) and functions $p_n(x)$. Identity (24) can be verified by expanding the solution of Eq. (8) around the edge of the Bloch band $\omega = \omega_{0,n}$. Utilizing these identities and (11), the Fredholm conditions for Eq. (21) finally lead to the following coupled nonlinear equations for the envelope functions A_1 and A_2:

$$\begin{aligned} D_1\frac{\partial^2 A_1}{\partial X^2} + D_2\frac{\partial^2 A_1}{\partial Y^2} + \eta A_1 + \sigma \left[\alpha|A_1|^2 A_1 + \beta\left(\bar{A}_1 A_2^2 + 2A_1|A_2|^2\right)\right. \\ \left. +\gamma\left(|A_2|^2 A_2 + \bar{A}_2 A_1^2 + 2A_2|A_1|^2\right)\right] = 0, \end{aligned} \tag{26}$$

$$\begin{aligned} D_2\frac{\partial^2 A_2}{\partial X^2} + D_1\frac{\partial^2 A_2}{\partial Y^2} + \eta A_2 + \sigma \left[\alpha|A_2|^2 A_2 + \beta\left(\bar{A}_2 A_1^2 + 2A_2|A_1|^2\right)\right. \\ \left. +\gamma\left(|A_1|^2 A_1 + \bar{A}_1 A_2^2 + 2A_1|A_2|^2\right)\right] = 0. \end{aligned} \tag{27}$$

Here

$$\alpha = \frac{\displaystyle\int_0^{2L}\int_0^{2L} p_1^4(x)p_2^4(y)\, dxdy}{\displaystyle\int_0^{2L}\int_0^{2L} p_1^2(x)p_2^2(y)\, dxdy}, \tag{28}$$

$$\beta = \frac{\displaystyle\int_0^{2L}\int_0^{2L} p_1^2(x)p_2^2(x)p_1^2(y)p_2^2(y)\, dxdy}{\displaystyle\int_0^{2L}\int_0^{2L} p_1^2(x)p_2^2(y)\, dxdy}, \tag{29}$$

and

$$\gamma = \frac{\int_0^{2L}\int_0^{2L} p_1^3(x)p_2(x)p_2^3(y)p_1(y)\,dxdy}{\int_0^{2L}\int_0^{2L} p_1^2(x)p_2^2(y)\,dxdy}. \tag{30}$$

Notice that α and β are always positive, but γ may be positive or negative.

The coefficients in Eqs. (26)-(27) can be readily determined from solutions of the 1D equation (8). In particular, at point C,

$$D_1 = 0.434845, \quad D_2 = 2.422196, \quad \alpha = 0.142814, \quad \beta = 0.032511, \quad \gamma = 0; \tag{31}$$

at point D,

$$D_1 = 0.586799, \quad D_2 = 13.264815, \quad \alpha = 0.086031, \quad \beta = 0.029655, \quad \gamma = 0; \tag{32}$$

and at point E,

$$D_1 = 0.434845,\ D_2 = 15.793172,\ \alpha = 0.971951,\ \beta = 0.162160,\ \gamma = -0.054081. \tag{33}$$

It is noted that near band edges where a single Bloch mode exists (such as points A and B in Fig. 3), the envelope equation for this single Bloch mode can be more easily derived. In this case, this single Bloch mode must be of the form $p(x;\omega_n)p(y;\omega_n)$, where ω_n is a band edge in the 1D problem (8). The resulting envelope equation for this Bloch mode is

$$D_1\left(\frac{\partial^2 A_1}{\partial X^2} + \frac{\partial^2 A_1}{\partial Y^2}\right) + \eta A_1 + \sigma\alpha_0|A_1|^2 A_1 = 0, \tag{34}$$

where

$$D_1 = \frac{1}{2}\frac{d^2\omega}{dk^2}\bigg|_{\omega=\omega_n}, \qquad \alpha_0 = \frac{\int_0^{2L}\int_0^{2L} p_1^4(x)p_1^4(y)\,dxdy}{\int_0^{2L}\int_0^{2L} p_1^2(x)p_1^2(y)\,dxdy}, \tag{35}$$

and $p_1(x) = p(x;\omega_n)$.

From the above asymptotic solutions, we can calculate the power of the corresponding composite solitary wave as $\epsilon \to 0$ (i.e. on the band edge). Details will be omitted here.

5. Solutions of the coupled envelope equations

Envelope equations (26)-(27) are the key results of this article. They have important consequences. First, they show that solitary waves are possible only when $\eta D_1 < 0, \eta D_2 < 0$. In this case, μ lies in the bandgap of the linear system as expected (see Eq. (13). Second, they show that solitary waves exist only when the dispersion coefficients D_1, D_2 and the nonlinearity coefficient σ are of the same sign. For instance, at point C in Fig. 3 where $D_1 > 0, D_2 > 0$, solitary waves exist only when

$\sigma > 0$, i.e., for focusing nonlinearity, not for defocusing nonlinearity. The situation is opposite at point D.

Below we study solutions of envelope equations (26)-(27). This system allows various reductions. If $\gamma = 0$, it allows the following three simple reductions:

(a) $A_1 > 0$, $A_2 = 0$, or $A_1 = 0, A_2 \neq 0$. In this case, the solution is a single Bloch-wave envelope solution.

(b) $A_1 > 0$, $A_2 > 0$. In this case, the solution is a composite real-valued envelope state. Note that the $A_1 > 0, A_2 < 0$ solution, or $A_1 < 0, A_2 > 0$ solution, or $A_1 < 0, A_2 < 0$ solution, or A_1, A_2 being both purely imaginary solution, is equivalent to the $A_1 > 0, A_2 > 0$ solution in (26)-(27), and leads to the equivalent solitary waves in the original system (4).

(c) $A_1 > 0$, $A_2 = i\hat{A}_2$, $\hat{A}_2 > 0$. In this case, the solution is a composite complex-valued envelope state. Note that other solutions of A_1 purely imaginary and A_2 real are equivalent to this real A_1 and purely imaginary A_2 solution.

If $\gamma \neq 0$, however, the reductions are quite different. For instance, the first and third reductions of case $\gamma = 0$ no longer hold. In this case, the following two reductions are allowed:

(a) $A_1 > 0$, $A_2 > 0$. In this case, the solution is a composite real-valued envelope state;

(b) $A_1 > 0, A_2 < 0$. In this case, the solution is another composite real-valued envelope state *different* from the $A_1 > 0$, $A_2 > 0$ reduction.

It is note-worthy that at band edges with $\gamma \neq 0$, the single Bloch-wave envelope reduction of $A_1 \neq 0$, $A_2 = 0$ or $A_1 = 0, A_2 \neq 0$ is not possible. Physically, this is due to a resonance between the two Bloch modes, which prevents the existence of a single Bloch mode envelope solution. For instance, at point E in Fig. 3 where the two Bloch solutions are $p(x;\omega_1)p(y;\omega_5)$ and $p(y;\omega_1)p(x;\omega_5)$, both $p(x;\omega_1)$ and $p(x;\omega_5)$ are symmetric in x and have period π. Thus these two modes are in resonance. At points where $\gamma = 0$ (such as point C and D in Fig. 3), the two Bloch solutions are not in resonance due to different symmetries, thus single Bloch-wave reduction is possible there.

To illustrate the composite solitary waves admitted by Eqs. (26)-(27), we consider points C and D in Fig. 3, where $\gamma = 0$. We look for the third reduction discussed above, i.e., $A_1 > 0$, $A_2 = i\hat{A}_2$, $\hat{A}_2 > 0$. In this case, the envelope solutions A_1 and $\hat{A}_2$ near points C and D with $\epsilon = 0.2$ are displayed in Figs. 5 and 6 respectively. At point C, $\sigma = 1$ (focusing nonlinearity), and $\eta = -1$; while at point D, $\sigma = -1$ (defocusing nonlinearity), and $\eta = 1$.

It should be noted that even though the envelope equations (26)-(27) are translation-invariant along the X and Y directions, the original equation (4) does not allow that invariance due to the potential term. Hence envelopes A_1, A_2 can not be placed arbitrarily relative to the periodic potential. In the 1D case, it has been shown that the envelope solution can only be placed at two special locations of the potential [12]. In the present 2D case, we can show that envelopes (A_1, A_2) can only be placed at four special positions relative to the periodic potentials. Specifically, the centers of these envelopes must be at $(x, y) = (0,0), (0, \pi/2), (\pi/2, 0)$ or $(\pi/2, \pi/2)$, hence four different solitary waves can be obtained. Of course, these center positions can also be shifted by multiple periods π along either of the x and y directions, but the resulting solutions are equivalent to the four mentioned above.

When envelope solutions (A_1, A_2) of Eqs. (26)-(27) are substituted into the perturbation series (12), solutions of the original system (4) will be analytically obtained. To illustrate, we take the envelope solutions displayed in Figs. 5 and 6 for points C and D, and let them be centered at the origin $(x, y) = (0,0)$. Substituting these envelopes into the expansion (12), the leading-order solutions of Eq. (4) near points C and D are displayed in Fig. 7. We see that these solutions have interesting amplitude and phase structures. These structures have many common features. First, the amplitude fields of both solutions are dominant along the x and y directions, forming a cross pattern. Second, at the center of each lattice, i.e. points $x = m\pi, y = n\pi$ with m, n being integers, the amplitudes are zero in both cases. Around each lattice center, the phase increases or decreases by 2π. Thus the solution around each lattice center has a vortex-cell structure. Because of this, we can call these solutions vortex-array solitons. Differences between these two solutions are also apparent. One difference is that, at point D, the whole field is divided into disconnected cells. But at point C, only the outer field seems divided into disconnected cells; the inner field is totally connected. Another difference is that, at point D, each cell is either a vortex ring or dipole. At point C, however, the cells look quite different.

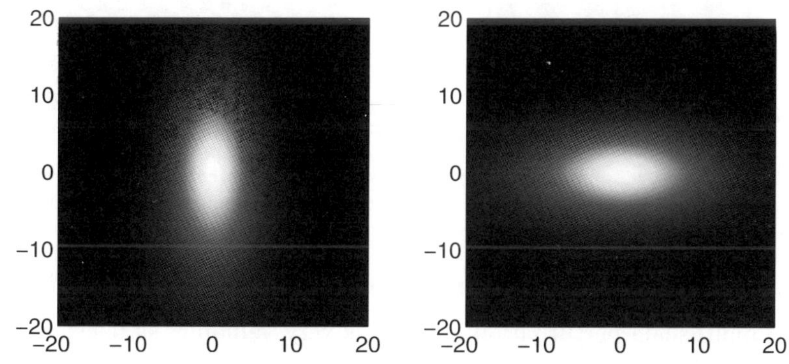

Figure 5. Envelop solutions A_1 (left) and $\hat{A}_2$ (right) near point C with $\epsilon = 0.2$.

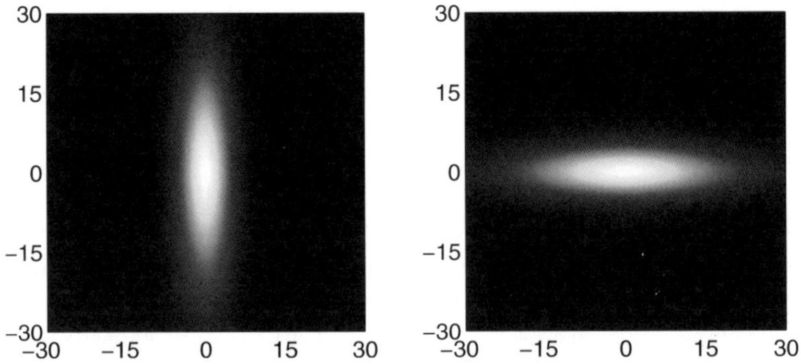

Figure 6. Envelop solutions A_1 (left) and $\hat{A}_2$ (right) near point D with $\epsilon = 0.2$.

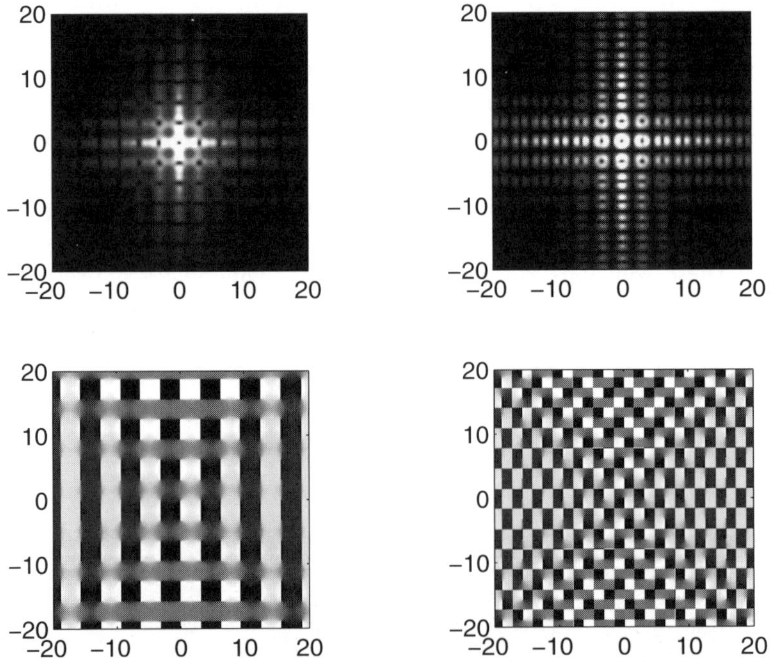

Figure 7. The leading-order analytical solutions near points C (left column) and D (right column) when $\epsilon = 0.2$. The top rows are amplitude plots, and the bottom rows are phase plots.

6. Numerical solutions of solitary waves at arbitrary amplitudes

The above multi-scale perturbation analysis is very valuable, as it clearly predicts various types of low-amplitude composite solitary waves near edges of Bloch bands. As the propagation constant μ moves away from these band edges, these solutions become more localized, and their amplitudes become higher. In such cases, the perturbation analysis starts to break down, and solutions need to be computed numerically. In this section, we numerically determine whole families of composite solitary-wave solutions bifurcating from edges of Bloch bands. The numerical method we use is the modified squared-operator iteration method described in [20]. In these numerical computations, the above analytical solutions from the perturbation analysis are very important, as they are the starting point of our iteration scheme.

For illustration purpose, we present the families of solutions bifurcating from the vortex-array solitons of Fig. 7. The power curves of these solution families are shown in Fig. 8. Both curves have a power threshold, below which the solutions do not exist. As μ approaches the band edges, the powers of the C-family (left) and D-family (right) approach 10.4254 and 19.5470 respectively. Two solutions on each family (marked in the power curves of Fig. 8) are displayed in Figs. 9 and 10. In both figures, the left solution is close to the band edge, while the right solution is deep inside the band gap. As expected, when the solution is close to the band edge, its amplitude is low, and it is similar to the analytical solution shown in Fig. 7. This is a partial confirmation of our asymptotic analysis in the previous section. Deep inside the band gap, however, the solutions are very localized, and their profiles look quite different from the low-amplitude solutions. The features of these localized solutions can not be gleamed entirely from the analytical solutions, thus their numerical computations are necessary and helpful. The vortex-array soliton in the right column of Fig. 9 corresponds to the higher-band vortex observed in [17] (where nonlinearity is of focusing type). The vortex-array soliton in the right column of Fig. 10 for defocusing nonlinearity has never been reported before in the literature.

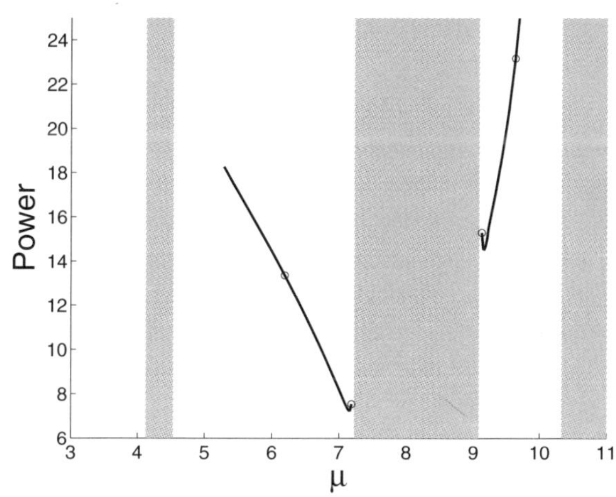

Figure 8. Power curves of composite vortex-array solitons bifurcating from points C (left) and D (right) of Fig. 3. The circle points are where we plot the numerical solutions in the following figures. In the left band gap, the circle points are 1.04 and 0.04 from the band edge. In the right band gap, the circle points are 0.54 and 0.04 from the band edge.

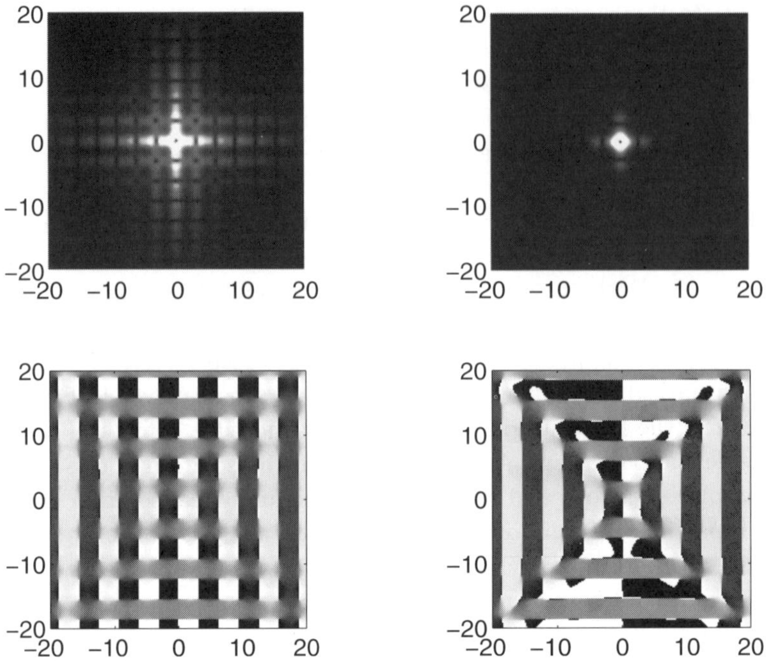

Figure 9. The amplitude (top) and phase (bottom) structures of two vortex-array solitons bifurcating from point C of Fig. 3 for focusing nonlinearity. The propagation constants of these two solutions are 7.1891 (left) and 6.1891 (right), as marked in Fig. 8.

7. Conclusion

In this paper, we obtained new solitary-wave structures in two-dimensional periodic media both analytically and numerically. These new structures bifurcate from the edges of Bloch bands with two linearly independent Bloch modes. Using perturbation methods, we derived the coupled nonlinear envelope equations for these composite solitons near the band edges. These envelope equations admit solutions which give rise to new soliton structures such as vortex-array solitons. Using numerical methods, we also computed these composite solitons directly both near and further away from the band edges. The numerical results are in full agreement with analytical ones near band edges.

Acknowledgments

This work was partially supported by the U.S. Air Force Office of Scientific Research under grant USAF 9550-05-1-0379.

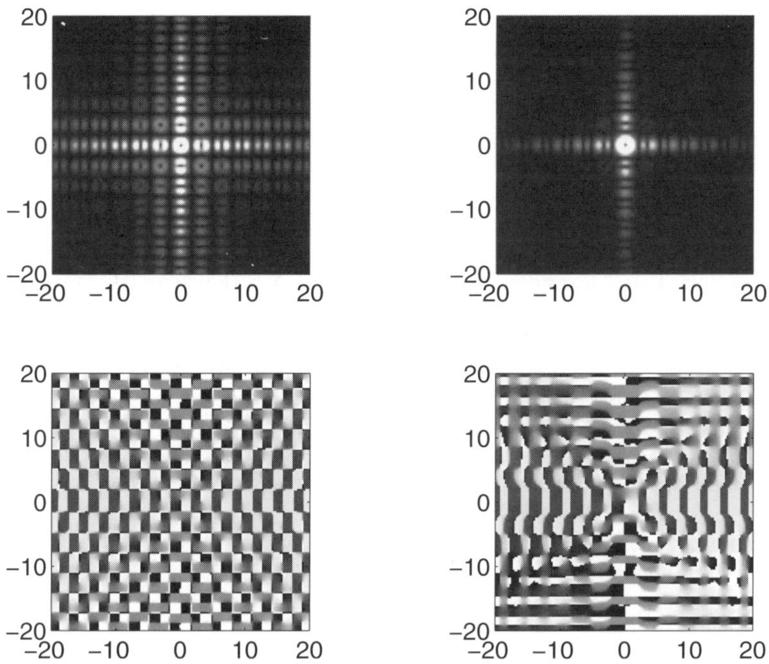

Figure 10. The amplitude (top) and phase (bottom) structures of two vortex-array solitons bifurcating from point D of Fig. 3 for defocusing nonlinearity. The propagation constants of these two solutions are 9.1210 (left) and 9.6210 (right), as marked in Fig. 8.

References

1. J.D. Joannopoulos, R.D. Meade, and J.N. Winn, *Photonic Crystals: Molding the Flow of Light*, Princeton University Press, 1995.
2. P. Russell, "Photonic Crystal Fibers", Science, 299, pp. 358 - 362 (2003).
3. H. S. Eisenberg, Y. Silberberg, R. Morandotti, A. R. Boyd, and J. S. Aitchison, Discrete Spatial Optical Solitons in Waveguide Arrays, Phys. Rev. Lett. 81, 3383-3386 (1998).
4. J.W Fleischer, M. Segev, N.K Efremidis, and D.N Christodoulides, "Observation of two-dimensional discrete solitons in optically induced nonlinear photonic lattices", Nature 422, 147 (2003).
5. H. Martin, E.D. Eugenieva, Z. Chen and D.N. Christodoulides, Discrete solitons and soliton-induced dislocations in partially-coherent photonic lattices, Phys. Rev. Lett. 92, 123902 (2004).
6. R. Iwanow, R. Schiek, G. I. Stegeman, T. Pertsch, F. Lederer, Y. Min, and W. Sohler, Observation of Discrete Quadratic Solitons, Phys. Rev. Lett. 93, 113902(2004).
7. F. Fedele, J. Yang, and Z. Chen, "Defect modes in one-dimensional photonic latices." Opt. Lett. 30, 1506 (2005).
8. I. Makasyuk, Z. Chen and J. Yang, "Bandgap guidance in optically-induced photonic lattices with a negative defect", Phys. Rev. Lett. 96, 223903 (2006).
9. F. Dalfovo, S. Giorgini, L. P. Pitaevskii, and S. Stringari, "Theory of Bose-Einstein condensation in trapped gases", Rev. Mod. Phys. 71, 463 (1999).
10. J.W. Fleischer, T. Carmon, M. Segev, N.K. Efremidis, and D.N. Christodoulides, Observation of Discrete Solitons in Optically Induced Real Time Waveguide Arrays,

Phys. Rev. Lett. 90, 023902 (2003).
11. D. Neshev, E. Ostrovskaya, Yu.S. Kivshar, W. Krolikowski, Spatial solitons in optically induced gratings, Opt. Lett. 28, 710 (2003).
12. D. E. Pelinovsky, A. A. Sukhorukov, and Y. S. Kivshar, "Bifurcations and stability of gap solitons in periodic potentials", Phys. Rev. E 70, 036618 (2004).
13. B. A. Malomed and P. G. Kevrekidis, Discrete vortex solitons, Phys. Rev. E 64, 026601 (2001).
14. J. Yang and Z.H. Musslimani, Fundamental and vortex solitons in a two-dimensional optical lattice. Opt. Lett. 23, 2094 (2003).
15. D.N. Neshev, T.J. Alexander, E.A. Ostrovskaya, Y.S. Kivshar, H. Martin, Z. Chen, Observation of Discrete Vortex Solitons in Optically Induced Photonic Lattices. Phys. Rev. Lett. 92, 123903 (2004).
16. J.W. Fleischer, G. Bartal, O. Cohen, O. Manela, M. Segev, J. Hudock, D.N. Christodoulides, Observation of vortex-ring discrete solitons in 2D photonic lattices. Phys. Rev. Lett. 92, 123904 (2004).
17. G. Bartal, O. Manela, O. Cohen, J.W. Fleischer, and M. Segev, "Observation of Second-Band Vortex Solitons in 2D Photonic Lattices", Phys. Rev. Lett. 95, 053904 (2005).
18. I. Makasyuk, Z. Chen and J. Yang, "Bandgap guidance in optically-induced photonic lattices with a negative defect", Phys. Rev. Lett. 96, 223903 (2006).
19. J. Yang, Stability of vortex solitons in a photorefractive optical lattice. New Journal of Physics 6, 47 (2004).
20. J. Yang and T.I. Lakoba, "Squared-operator iteration methods for solitary waves in general nonlinear wave equations." To appear in Stud. Appl. Math.